# UNDERGRADUATE
# CONVEXITY
## Problems and Solutions

# UNDERGRADUATE CONVEXITY
## Problems and Solutions

**Mikkel Slot Nielsen**
**Victor Ulrich Rohde**

Aarhus University, Denmark

## World Scientific

NEW JERSEY · LONDON · SINGAPORE · BEIJING · SHANGHAI · HONG KONG · TAIPEI · CHENNAI · TOKYO

*Published by*

World Scientific Publishing Co. Pte. Ltd.

5 Toh Tuck Link, Singapore 596224

*USA office:* 27 Warren Street, Suite 401-402, Hackensack, NJ 07601

*UK office:* 57 Shelton Street, Covent Garden, London WC2H 9HE

**Library of Congress Cataloging-in-Publication Data**
Names: Nielsen, Mikkel Slot. | Rohde, Victor Ulrich. | Lauritzen, Niels,
  1964–   Undergraduate convexity.
Title: Undergraduate convexity : problems and solutions / by Mikkel Slot Nielsen
  (Aarhus University, Denmark), Victor Ulrich Rohde (Aarhus University, Denmark).
Description: New Jersey : World Scientific, 2016. | The answer book to exercises found in
  Undergraduate convexity : from Fourier and Motzkin to Kuhn and Tucker
  (Singapore ;  Hackensack, NJ : World Scientific, c2013).
Identifiers: LCCN 2016025415| ISBN 9789813146211 (hardcover : alk. paper) |
  ISBN 9789813143647 (pbk. : alk. paper)
Subjects: LCSH: Convex functions--Examinations, questions, etc. |
  Convex domains--Examinations, questions, etc.
Classification: LCC QA331.5 .L295 Suppl. 2016 | DDC 515/.882--dc23
LC record available at https://lccn.loc.gov/2016025415

**British Library Cataloguing-in-Publication Data**
A catalogue record for this book is available from the British Library.

Printed in Singapore

# Preface

Our aim has been to provide a solutions manual to the exercises of Niels Lauritzen's book 'Undergraduate Convexity: From Fourier and Motzkin to Kuhn and Tucker', 2013 (from this point referred to as U.C.). This is partly a composition of solutions made during our periods as teaching assistants in the courses *Konvekse Mængder* (convex sets) and *Konvekse Funktioner* (convex functions), but the final manual is extensive in the sense that one will find solutions to all exercises together with detailed explanations and illustrations. However, to avoid too many repetitions, the first exercises in a given chapter are solved with more details than the last exercises in the chapter. Another reason is that the reader becomes more familiar with the different concepts later on and hence, needs less amount of explanations.

Each chapter in this manual starts with a minor introduction to the associated chapter in U.C. It will summarize key results, measured both in theoretical significance and in the importance of the solutions, together with relevant definitions. Naturally, one will find references to U.C., and these will be stated explicitly whenever there is ambiguity. For instance, [U.C., (9.6)] refers to (9.6) in U.C. In addition to this introducing section, the solution sections consist both of exercise descriptions and the actual solutions. The idea behind this construction is that the reader can use this manual without having the main book next to them. However, the manual cannot in any way substitute U.C. since the solutions are built on the assumption that the reader has been reading the book (or at least the associated chapter) beforehand.

Finally, the chapters are written independently of each other as much as possible and as a consequence, one can go directly to the chapter of interest without lacking any knowledge in order to understand the solutions.

Aarhus, March 2016

# Acknowledgments

We sincerely appreciate the contributions of Cecilie Marie Løchte Jørgensen in completing the solutions to the exercises. Additionally, Ole Agersnap has come up with several inputs, and for this we are very grateful.

Niels Lauritzen has been a rich source of inspiration both through his book and in meetings, and he has made an abundance of good suggestions for improving the manual. All of this is and has been highly appreciated.

Finally, we are very thankful to Lars 'daleif' Madsen for his crucial assistance in layout, good conventions, and many other LaTeX related issues.

Acknowledgments

# Contents

# Chapter 1

# Fourier-Motzkin elimination

## 1.1 Introduction

Just like its Gaussian counterpart, Fourier-Motzkin elimination quickly becomes very natural with a bit of practice. The exercises below focus exactly on this practice, and the reader will hopefully feel confident in working with inequalities after carefully going through the material. Some exercises have a more applied flavor and may also serve as a motivation for reducing a system of inequalities.

Being cumbersome, space consuming, and slightly trivial to rewrite the Fourier-Motzkin elimination in detail in every solution, the later solutions become more concise; so, feeling a bit lost may be resolved by simply going back a couple of solutions.

**Proposition 1.1.** *Let* $\alpha_1, \ldots, \alpha_r, \beta_1, \ldots, \beta_s \in \mathbb{R}$. *Then*

$$\max\{\alpha_1, \ldots, \alpha_r\} \leq \min\{\beta_1, \ldots, \beta_s\}$$

*if and only if* $\alpha_i \leq \beta_j$ *for every* $i, j$ *with* $1 \leq i \leq r$ *and* $1 \leq j \leq s$:

$$\alpha_1 \leq \beta_1 \ldots \alpha_1 \leq \beta_s$$
$$\vdots \qquad \ddots \qquad \vdots$$
$$\alpha_r \leq \beta_1 \ldots \alpha_r \leq \beta_s.$$

**Definition 1.4.** The subset

$$P = \left\{ \begin{pmatrix} x_1 \\ \vdots \\ x_n \end{pmatrix} \in \mathbb{R}^n \left| \begin{array}{c} a_{11}x_1 + \cdots + a_{1n}x_n \leq b_1 \\ \vdots \\ a_{m1}x_1 + \cdots + a_{mn}x_n \leq b_m \end{array} \right. \right\} \subseteq \mathbb{R}^n$$

of solutions to a system

$$a_{11}x_1 + \cdots + a_{1n}x_n \leq b_1$$
$$\vdots$$
$$a_{m1}x_1 + \cdots + a_{mn}x_n \leq b_m$$

of finitely many linear inequalities (here $a_{ij}$ and $b_i$ are real numbers) is called a *polyhedron.*

Whilst Proposition 1.1 is very useful for computation, the following result is more of theoretical interest.

**Theorem 1.6.** *Consider the projection* $\pi : \mathbb{R}^n \to \mathbb{R}^{n-1}$ *given by*

$$\pi(x_1, \ldots, x_n) = (x_2, \ldots, x_n).$$

*If* $P \subseteq \mathbb{R}^n$ *is a polyhedron, then*

$$\pi(P) = \{(x_2, \ldots, x_n) \mid \exists\, x_1 \in \mathbb{R} : (x_1, x_2, \ldots, x_n) \in P\} \subseteq \mathbb{R}^{n-1}$$

*is a polyhedron.*

## 1.2   Exercises and solutions

**Exercise 1.1.** Sketch the set of solutions to the system

$$\begin{aligned}
2x + y &\geq 2 \\
3x + y &\leq 9 \\
-x + 2y &\leq 4 \\
y &\geq 0
\end{aligned} \tag{1.1}$$

of linear inequalities. Carry out the elimination procedure for (1.1) as illustrated in §1.1.

**Solution 1.1.** The set of solutions is sketched in Figure 1.1. To carry out the elimination procedure we start by isolating $x$ which gives the new system

$$\begin{aligned}
1 - \tfrac{1}{2}y &\leq x \\
x &\leq 3 - \tfrac{1}{3}y \\
-4 + 2y &\leq x \\
0 &\leq y.
\end{aligned}$$

This is, according to Proposition 1.1, equivalent to the system

$$\max\{1 - \tfrac{1}{2}y, 2y - 4\} \le x \le 3 - \tfrac{1}{3}y$$
$$0 \le y.$$

There exists $x$ such that $(x, y)$ is a solution if and only if $y$ satisfies

$$\max\{1 - \tfrac{1}{2}y, 2y - 4\} \le 3 - \tfrac{1}{3}y \qquad (1.2)$$
$$0 \le y.$$

By another use of Proposition 1.1 we know that $y$ is a solution to (1.2) if and only if $y$ is a solution to

$$1 - \tfrac{1}{2}y \le 3 - \tfrac{1}{3}y$$
$$2y - 4 \le 3 - \tfrac{1}{3}y$$
$$0 \le y.$$

This system is equivalent to

$$-12 \le y$$
$$y \le 3$$
$$0 \le y$$

which means that a solution satisfies $y \in [0, 3]$. In conclusion, we have $(x, y) \in \mathbb{R}^2$ is a solution if and only if $y \in [0, 3]$ and

$$\max\left\{1 - \tfrac{1}{2}y, 2y - 4\right\} \le x \le 3 - \tfrac{1}{3}y.$$

$$\star \quad \star \quad \star$$

**Exercise 1.2.** Let

$$P = \left\{(x, y, z) \in \mathbb{R}^3 \;\middle|\; \begin{array}{rrrr} -x - & y - & z \le 0 \\ 3x - & y - & z \le 1 \\ -x + & 3y - & z \le 2 \\ -x - & y + & 3z \le 3 \end{array}\right\}$$

and $\pi : \mathbb{R}^3 \to \mathbb{R}^2$ be given by $\pi(x, y, z) = (y, z)$.

   (i) Compute $\pi(P)$ as a polyhedron i.e., as the solutions to a set of linear inequalities in $y$ and $z$.
   (ii) Compute $\eta(P)$, where $\eta : \mathbb{R}^3 \to \mathbb{R}$ is given by $\eta(x, y, z) = x$.
   (iii) How many integral points does $P$ contain i.e., how many elements are in the set $P \cap \mathbb{Z}^3$?

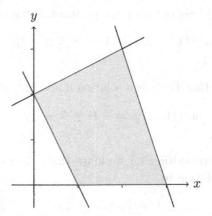

**Figure 1.1:** The set of solutions

**Solution 1.2.**     (i) From the description above we know that $P$ is the set of $(x, y, z) \in \mathbb{R}^3$ where

$$
\begin{aligned}
-y - z &\le x \\
x &\le \tfrac{1}{3} + \tfrac{1}{3}y + \tfrac{1}{3}z \\
3y - z - 2 &\le x \\
-y + 3z - 3 &\le x.
\end{aligned}
$$

From Proposition 1.1 we have the equivalent system

$$
\max\{-y - z, 3y - z - 2, -y + 3z - 3\} \le x \le \tfrac{1}{3} + \tfrac{1}{3}y + \tfrac{1}{3}z. \quad (1.3)
$$

Another use of Proposition 1.1 implies that (1.3) is solvable in $x$ if and only if

$$
\begin{aligned}
-y - z &\le \tfrac{1}{3} + \tfrac{1}{3}y + \tfrac{1}{3}z \\
3y - z - 2 &\le \tfrac{1}{3} + \tfrac{1}{3}y + \tfrac{1}{3}z \\
-y + 3z - 3 &\le \tfrac{1}{3} + \tfrac{1}{3}y + \tfrac{1}{3}z.
\end{aligned}
$$

Consequently,

$$
\pi(P) = \left\{ (y, z) \in \mathbb{R}^2 \;\middle|\; \begin{array}{r} -4y - 4z \le 1 \\ 8y - 4z \le 7 \\ -4y + 8z \le 10 \end{array} \right\}.
$$

(ii) We start by eliminating $z$. Therefore, we write $P$ as the set of

$(x, y, z) \in \mathbb{R}^3$ with

$$
\begin{aligned}
-x - y &\leq z \\
3x - y - 1 &\leq z \\
-x + 3y - 2 &\leq z \\
z &\leq 1 + \tfrac{1}{3}x + \tfrac{1}{3}y
\end{aligned}
$$

and hence we must have

$$
\max\{-x - y, 3x - y - 1, -x + 3y - 2\} \leq 1 + \tfrac{1}{3}x + \tfrac{1}{3}y.
$$

By Proposition 1.1 the above may be written as

$$
\begin{aligned}
-x - \tfrac{3}{4} &\leq y \\
2x - \tfrac{3}{2} &\leq y \\
y &\leq \tfrac{1}{2}x + \tfrac{9}{8}.
\end{aligned}
$$

Eliminating $y$ in the same way results in $-5/4 \leq x \leq 7/4$.

(iii) From the conclusion of (ii) we know that if $(x, y, z)$ has to be an integral point, $x \in \{-1, 0, 1\}$. For $x = 1$ the restriction on $y$ is

$$
\tfrac{1}{2} = \max\{-1 - \tfrac{3}{4}, 2 - \tfrac{3}{2}\} \leq y \leq \tfrac{1}{2} + \tfrac{9}{8} = \tfrac{13}{8}.
$$

This shows that if $x = 1$ then $y = 1$, when $y$ is an integer. Setting $x = y = 1$ we have that $z$ must satisfy

$$
1 = \max\{-1 - 1, 3 - 1 - 1, -1 + 3 - 2\} \leq z \leq 1 + \tfrac{1}{3} + \tfrac{1}{3} = \tfrac{5}{3}.
$$

Thus, the first integral point is $(1, 1, 1)$. Similar calculations lead to the final conclusion

$$
P \cap \mathbb{Z}^3 = \{(1, 1, 1), (0, 0, 0), (0, 0, 1), (0, 1, 1)\}
$$

giving a total of four integral points.

$$\star \quad \star \quad \star$$

**Exercise 1.3.** Find all solutions $x, y, z \in \mathbb{Z}$ to the linear inequalities

$$
\begin{aligned}
-x + y - z &\leq 0 \\
-y + z &\leq 0 \\
-z &\leq 0 \\
x \quad\quad - z &\leq 1 \\
y \quad\quad &\leq 1 \\
z &\leq 1
\end{aligned}
$$

by using Fourier-Motzkin elimination.

**Solution 1.3.** By Proposition 1.1 the system is equivalent to

$$\max\{-x+y, 0, x-1\} \le z \le \min\{y, 1\}$$
$$y \le 1$$

which has a solution in $z$ if and only if

$$0 \le x$$
$$y - 1 \le x$$
$$x \le y + 1$$
$$x \le 2$$
$$0 \le y \le 1.$$

In short, $\max\{0, y-1\} \le x \le \min\{y+1, 2\}$ and $0 \le y \le 1$. By use of Proposition 1.1, we find that $\max\{0, y-1\} \le \min\{y+1, 2\}$ is always satisfied for $0 \le y \le 1$ making the restriction redundant. For $y = 0$ we find that $0 = \max\{0, -1\} \le x \le \min\{1, 2\} = 1$. Thus, when $y = 0$, $x$ is either 0 or 1. For $(0, 0, z)$ to be a solution we must have $0 = \max\{0, 0, -1\} \le z \le \min\{0, 1\} = 0$ making $(0, 0, 0)$ feasible. With similar calculations we find a total of six integral points: $(0, 0, 0), (1, 0, 0), (0, 1, 1), (1, 1, 0), (1, 1, 1)$, and $(2, 1, 1)$.

$$\star \quad \star \quad \star$$

**Exercise 1.4.** Does the system

$$2x - 3y + z \le -2$$
$$x + 3y + z \le -3$$
$$-2x - 3y + z \le -2$$
$$-x - 3y - 3z \le 1$$
$$-2x - y + 3z \le 3$$

of linear inequalities have a solution $x, y, z \in \mathbb{R}$?

**Solution 1.4.** Let us define the polyhedron

$$P = \left\{ (x, y, z) \in \mathbb{R}^3 \middle| \begin{pmatrix} 2 & -3 & 1 \\ 1 & 3 & 1 \\ -2 & -3 & 1 \\ -1 & -3 & -3 \\ -2 & -1 & 3 \end{pmatrix} \begin{pmatrix} x \\ y \\ z \end{pmatrix} \le \begin{pmatrix} -2 \\ -3 \\ -2 \\ 1 \\ 3 \end{pmatrix} \right\}.$$

The question is then if $P = \emptyset$ or not. Defining $\pi : \mathbb{R}^3 \to \mathbb{R}$ as the projection onto the $z$-coordinate, an equivalent question is whether $\pi(P) = \emptyset$ or not, which we can answer by explicitly computing $\pi(P)$ with the help of Fourier-Motzkin elimination (first, we will project a solution $(x, y, z)$ onto $(y, z)$ and second, project $(y, z)$ onto $z$). The inequalities can be written equivalently as

$$
\begin{aligned}
x &\leq -1 + \tfrac{3}{2}y - \tfrac{1}{2}z \\
x &\leq -3 - 3y - z \\
1 - \tfrac{3}{2}y + \tfrac{1}{2}z &\leq x \\
-1 - 3y - 3z &\leq x \\
-\tfrac{3}{2} - \tfrac{1}{2}y + \tfrac{3}{2}z &\leq x,
\end{aligned}
$$

which is solvable in $x$ if and only if

$$
\max\left\{1 - \tfrac{3}{2}y + \tfrac{1}{2}z, -1 - 3y - 3z, -\tfrac{3}{2} - \tfrac{1}{2}y + \tfrac{3}{2}z\right\} \\
\leq \min\left\{-1 + \tfrac{3}{2}y - \tfrac{1}{2}z, -3 - 3y - z\right\}. \tag{1.4}
$$

By repeated use of Proposition 1.1 and rearranging terms we find the equivalent system

$$
\max\{\tfrac{3}{2} + \tfrac{1}{2}z, -\tfrac{5}{9}z, -\tfrac{1}{4} + z\} \leq y \leq \min\{-\tfrac{8}{3} - z, -\tfrac{3}{5} - z\} \tag{1.5}
$$

and $z \geq 1$. Necessary for (1.5) to hold is, for instance, that $-1/4 + z \leq -8/3 - z$, which is the same as $z \leq -29/24$. Since this inequality together with $z \geq 1$ cannot hold we conclude that $\pi(P) = \emptyset$.

$$\star \quad \star \quad \star$$

**Exercise 1.5.** Let $P \subseteq \mathbb{R}^n$ be a polyhedron and $c \in \mathbb{R}^n$. Define the polyhedron $P' \subseteq \mathbb{R}^{n+1}$ by

$$
P' = \{\left(\begin{smallmatrix} z \\ x \end{smallmatrix}\right) \in \mathbb{R}^{n+1} \mid c^t x = z, \; x \in P, \; z \in \mathbb{R}\}.
$$

  (i) How does this setup relate to Example 1.2?
 (ii) Show how projection onto the $z$-coordinate (and Fourier-Motzkin elimination) in $P'$ can be used to solve the linear optimization problem of finding $x \in P$, such that $c^t x$ is minimal (or proving that such an $x$ does not exist).
(iii) Let $P$ denote the polyhedron from Exercise 1.2. You can see that

$$
(0, 0, 0), \; (-1, \tfrac{1}{2}, \tfrac{1}{2}) \in P
$$

have values $0$ and $-1$ on their first coordinates, but what is the minimal first coordinate of a point in $P$?

**Solution 1.5.**     (i)  In Example 1.2 we tried to maximize $x+y$ given some restrictions on $(x, y) \in \mathbb{R}^2$. The set of solutions to these restrictions is given by the polyhedron $P \subseteq \mathbb{R}^2$ where

$$P = \left\{ (x,y) \in \mathbb{R}^2 \,\middle|\, \begin{pmatrix} 1 & 2 \\ 2 & 1 \\ -1 & 0 \\ 0 & -1 \end{pmatrix} \begin{pmatrix} x \\ y \end{pmatrix} \leq \begin{pmatrix} 3 \\ 3 \\ 0 \\ 0 \end{pmatrix} \right\}.$$

Then we introduced $z$ as the value of the objective function of interest, which corresponds to $c = (1, 1)$, and defined

$$P' = \left\{ (x,y,z) \in \mathbb{R}^3 \,\middle|\, \begin{pmatrix} x \\ y \end{pmatrix} \in P, \ (1,1) \begin{pmatrix} x \\ y \end{pmatrix} = z \right\}.$$

With this setup we have that $(x, y, z) \in P'$ if and only if it solves the system at the top of [U.C., p. 9]. Subsequently, we found that the possible values of $z$, if $(x, y) \in P$, were given as the interval $[0, 2]$. In our setup it corresponds to deriving that $\eta(P') = [0, 2]$, where $\eta : \mathbb{R}^3 \to \mathbb{R}$ is given as the projection onto the $z$-coordinate.

(ii)  It is possible to generalize our discussion above. If we have given a polyhedron $P \subseteq \mathbb{R}^n$ and we are interested in optimizing (minimizing or maximizing) a linear functional $f : \mathbb{R}^n \to \mathbb{R}$ given by $f(x) = c^t x$, we can make use of $P'$. If we define the projection $\eta : \mathbb{R}^{n+1} \to \mathbb{R}$ by $\eta(x, z) = z$ we have that

$$\eta(P') = \{z \in \mathbb{R} | \ \exists x \in \mathbb{R}^n : (x, z) \in P'\} = \{z \in \mathbb{R} | \ \exists x \in P : c^t x = z\}.$$

Thus we can interpret $\eta(P')$ as the set of possible values of $f(x)$ provided that $x \in P$. It follows from Theorem 1.6 that $\eta(P')$ will be either a closed interval (possibly unbounded) or the empty set. Consequently, if we want to minimize $f(x)$ when $x \in P$, we just choose the smallest $z \in \eta(P')$ and if the set is empty or the interval is downward unbounded, we conclude that no such minimum exists. The set $\eta(P')$ is computed using Fourier-Motzkin elimination.

(iii)  Since we found in Exercise 1.2 that

$$\eta(P) = \{x \in \mathbb{R} | \ \exists (y, z) \in \mathbb{R}^2 : (x, y, z) \in P\} = \left[ -\tfrac{5}{4}, \tfrac{7}{4} \right],$$

it follows that the minimum first coordinate of a point in $P$ is $-5/4$.

$$\star \quad \star \quad \star$$

**Exercise 1.6.** Solve the problem appearing in Fourier's article ([U.C., Figure 1.6]) for $r = 1$ using Fourier-Motzkin elimination.

**Solution 1.6.** We need to find all $(x, y, z) \in \mathbb{R}^3$ such that

$$\max\{x, y, z\} \leq 2 \cdot \min\{x, y, z\}, \quad x + y + z = 1, \quad \text{and} \quad x, y, z \geq 0. \quad (1.6)$$

By Proposition 1.1, the inequalities in (1.6) may be written as

$$\begin{pmatrix} 1 & 0 & 0 \\ 1 & 0 & 0 \\ 1 & 0 & 0 \\ 0 & 1 & 0 \\ 0 & 1 & 0 \\ 0 & 1 & 0 \\ 0 & 0 & 1 \\ 0 & 0 & 1 \\ 0 & 0 & 1 \end{pmatrix} \begin{pmatrix} x \\ y \\ z \end{pmatrix} \leq 2 \begin{pmatrix} 1 & 0 & 0 \\ 0 & 1 & 0 \\ 0 & 0 & 1 \\ 1 & 0 & 0 \\ 0 & 1 & 0 \\ 0 & 0 & 1 \\ 1 & 0 & 0 \\ 0 & 1 & 0 \\ 0 & 0 & 1 \end{pmatrix} \begin{pmatrix} x \\ y \\ z \end{pmatrix}.$$

(Note that $x, y, z \geq 0$ are redundant restrictions.) We can now use Fourier-Motzkin elimination to see that the set of solutions to the problem is the points $(x, y, z) \in \mathbb{R}^3$ satisfying

$$\tfrac{1}{5} \leq z \leq \tfrac{1}{2}$$
$$\max\{\tfrac{1}{3} - \tfrac{1}{3}z, \ 1 - 3z, 0, \tfrac{1}{2}z\} \leq y \leq \min\{1 - z, \ \tfrac{2}{3} - \tfrac{2}{3}z, \ 2z, \ 1 - \tfrac{3}{2}z\}$$
$$x = 1 - y - z.$$

$$\star \quad \star \quad \star$$

**Exercise 1.7.** Let $P$ denote the set of $(x, y, z) \in \mathbb{R}^3$, satisfying

$$\begin{aligned} -2x + \ y + z &\leq 4 \\ x \qquad\quad &\geq 1 \\ y \quad\ &\geq 2 \\ z &\geq 3 \\ x - 2y + z &\leq 1 \\ 2x + 2y - z &\leq 5. \end{aligned}$$

(i) Prove that $P$ is bounded.
(ii) Find $(x, y, z) \in P$ with $z$ maximal. Is such a point unique?

**Solution 1.7.**    (i) We can answer this question by computing $\pi_1(P)$, $\pi_2(P)$, and $\pi_3(P)$, where $\pi_1, \pi_2, \pi_3 : \mathbb{R}^3 \to \mathbb{R}$ are the projections onto the $x$-, $y$-, and $z$-coordinate, respectively. By definition of the projection we have that

$$P \subseteq \bar{P} = \{(x, y, z) \in \mathbb{R}^3 \,|\, x \in \pi_1(P), \ y \in \pi_2(P), \ z \in \pi_3(P)\}$$

and hence if $\bar{P}$ is bounded, $P$ is bounded as well. As in former exercises we compute the projections of $P$ with the help of Fourier-Motzkin elimination. Starting by projecting onto the $x$-coordinate, we find that $\pi_1(P) = [0,2]$. Repeating the Fourier-Motzkin elimination starting from $P$ and projecting onto the $y$-coordinate gives that $\pi_2(P) = [2,3]$. Similarly, a projection onto the $z$-coordinate gives $\pi_3(P) = [3,5]$. Thus we have that $(x,y,z) \in \bar{P}$ implies $|(x,y,z)| \leq \sqrt{2^2 + 3^2 + 5^2}$, and $\bar{P}$ is indeed bounded.

**Remark.** Alternatively, instead of computing $\pi_2(P)$ and $\pi_3(P)$, one can use results from analysis to conclude that $P$ is bounded. Being more specific, we found that $\pi_1(P)$ is compact, and during the elimination procedure one will find

$$f_1(x) \leq y \leq g_1(x)$$
$$f_2(x,y) \leq z \leq g_2(x,y)$$

where $f_1$, $f_2$, $g_1$, and $g_2$ are continuous functions. Noting that we are considering $x$ and $(x,y)$ in compact sets, we get finite bounds on $y$ and $z$, respectively.

(ii) From (i) we see that the maximal value of $z$ if $(x,y,z) \in P$ is 5. From the computation of $\pi_3(P)$ we have that, for a given $z \in \pi_3(P)$, $y$ must satisfy

$$\max\{z - 2, \tfrac{1}{2}z, 2\} \leq y \leq \min\{3, \tfrac{1}{2}z + \tfrac{3}{2}\} \tag{1.7}$$

and after fixing such a $y$, $(x,y,z) \in P$ if and only if $x$ satisfies

$$\max\{\tfrac{1}{2}y + \tfrac{1}{2}z - 2, 1\} \leq x \leq \min\{2y - z + 1, -y + \tfrac{1}{2}z + \tfrac{5}{2}\}. \tag{1.8}$$

Choosing $z = 5$, we find from (1.7) that $y = 3$ and with $(y,z) = (3,5)$, (1.8) gives $x = 2$, so we conclude that $(2,3,5)$ is the unique point in $P$ that maximizes the value of the third coordinate.

$$\star \quad \star \quad \star$$

**Exercise 1.8.** A vitamin pill $P$ is produced using two ingredients $M_1$ and $M_2$. The pill needs to satisfy four constraints for the vital vitamins $V_1$ and $V_2$. It must contain at least 6 milligram and at most 15 milligram of $V_1$ and at least 5 milligram and at most 12 milligram of $V_2$. The ingredient $M_1$ contains 3 milligram of $V_1$ and 2 milligram of $V_2$ per gram. The ingredient $M_2$ contains 2 milligram of $V_1$ and 3 milligram of $V_2$ per gram:

|       | $V_1$ | $V_2$ |
| ----- | ----- | ----- |
| $M_1$ | 3     | 2     |
| $M_2$ | 2     | 3     |

Let $x$ denote the amount of $M_1$ and $y$ the amount of $M_2$ (measured in grams) in the production of a vitamin pill. Write down a system of linear inequalities in $x$ and $y$ describing the constraints above.

We want a vitamin pill of minimal weight satisfying the constraints. How many grams of $M_1$ and $M_2$ should we mix? Describe how Fourier-Motzkin elimination can be used in solving this problem.

**Solution 1.8.** From the description above we infer the inequalities

$$
\begin{aligned}
6 &\le 3x + 2y \le 15 \\
5 &\le 2x + 3y \le 12 \\
0 &\le x \\
0 &\le \phantom{x} y.
\end{aligned}
\tag{1.9}
$$

To find a vitamin pill of minimal weight we introduce $z = x + y$ and set out to minimize this $z$. First, substituting $x = z - y$ in (1.9) and then isolating $y$ yields

$$
\max\{3z - 15, 5 - 2z, 0\} \le y \le \min\{3z - 6, 12 - 2z, z\}.
\tag{1.10}
$$

There exists $y$ such that $(y, z)$ satisfies (1.10) if and only if

$$
\tfrac{11}{5} \le z \le \tfrac{27}{5}.
$$

We conclude that the minimum value is $z = 11/5$. From (1.10) we then find that $y = 3/5$ which gives $x = 11/5 - 3/5 = 8/5$. Thus, using 8/5 grams of $M_1$ and 3/5 grams of $M_2$ will minimize the weight of the vitamin pill.

$$\star \quad \star \quad \star$$

**Exercise 1.9.** Use Fourier-Motzkin elimination to compute the minimal value of

$$
x_1 + 2x_2 + 3x_3,
$$

when $x_1, x_2, x_3$ satisfy

$$
\begin{aligned}
x_1 - 2x_2 + x_3 &= 4 \\
-x_1 + 3x_2 \phantom{{} + x_3} &= 5
\end{aligned}
$$

and

$$
x_1 \ge 0, \quad x_2 \ge 0, \quad x_3 \ge 0.
$$

**Solution 1.9.** We start by letting $x_4 = x_1 + 2x_2 + 3x_3$. Isolating for $x_1$, substitution gives

$$-4x_2 - 2x_3 + x_4 = 4$$
$$5x_2 + 3x_3 - x_4 = 5$$
$$0 \leq -2x_2 - 3x_3 + x_4$$
$$0 \leq x_2$$
$$0 \leq x_3.$$

Solving for $x_2$ and $x_3$ in the two equalities above gives the inequalities

$$\tfrac{76}{3} \leq x_4 \leq 40,$$

and we may conclude that the minimum value for the problem is $x_4 = 76/3$. Working backwards through the previously found equalities then gives $x_1 = 0$, $x_2 = 5/3$ and $x_3 = 22/3$.

# Chapter 2

# Affine subspaces

## 2.1 Introduction

The exercises of this section are about the understanding of affine subspaces. This includes the special cases where we consider (genuine) subspaces, and here we focus on representing them as a set of solutions to a homogeneous system of equations. In addition, some exercises concern other kinds of affine subspaces and their representations as well as their link to subspaces. Another part of the exercises deals with the understanding of affine independence, and the geometric interpretation and link to linear independence will be in focus. Finally, there are a few exercises about affine maps and dimension of sets.

**Definition 2.2.** A non-empty subset $M \subseteq \mathbb{R}^d$ is called an *affine subspace* if

$$(1-t)u + tv \in M$$

for every $u, v \in M$ and every $t \in \mathbb{R}$. A map $f : \mathbb{R}^d \to \mathbb{R}^n$ is called an *affine map* if

$$f((1-t)u + tv) = (1-t)f(u) + tf(v)$$

for every $u, v \in \mathbb{R}^d$ and every $t \in \mathbb{R}$.

For $S \subseteq \mathbb{R}^d$ we define the affine hull of $S$ as

$$\text{aff}(S) := \{\lambda_1 v_1 + \cdots + \lambda_m v_m \mid m \geq 1, \ v_1, \ldots, v_m \in S, \ \lambda_1 + \cdots + \lambda_m = 1\}.$$

**Proposition 2.5.** *The affine hull, $\text{aff}(S)$, of a subset $S \subseteq \mathbb{R}^d$ is an affine subspace. It is the smallest affine subspace containing $S$.*

An alternative way of formulating the last part of Proposition 2.5 is that if $S \subseteq M$ and $M$ is an affine subspace, $\text{aff}(S) \subseteq M$.

13

In relation to the next result, we recall that a map $h : \mathbb{R}^d \to \mathbb{R}^n$ is linear if $h(\alpha x + \beta y) = \alpha h(x) + \beta h(y)$ for every $x, y \in \mathbb{R}^d$ and $\alpha, \beta \in \mathbb{R}$.

**Proposition 2.6.** *For an affine subspace $M \subseteq \mathbb{R}^d$, $W = \{u - v \mid u, v \in M\}$ is a subspace and*

$$M = \{x_0 + w \mid w \in W\} =: x_0 + W,$$

*for every $x_0 \in M$.*

*A subset $M \subseteq \mathbb{R}^d$ is an affine subspace if and only if it is the solution set to a system of linear equations.*

*If $h : \mathbb{R}^d \to \mathbb{R}^n$ is a linear map and $b \in \mathbb{R}^n$, then $f(x) = h(x) + b$ is an affine map. If $f : \mathbb{R}^d \to \mathbb{R}^n$ is an affine map, $h(x) = f(x) - f(0)$ is a linear map and $f(x) = h(x) + b$ with $b = f(0)$.*

**Proposition 2.9.** *Let $S = \{v_1, \ldots, v_m\} \subseteq \mathbb{R}^d$. Then $\mathrm{aff}(S) = v_1 + W$, where $W$ is the subspace spanned by $v_2 - v_1, \ldots, v_m - v_1$. The following conditions are equivalent.*

(1) *$S$ is affinely independent.*
(2) *$v_2 - v_1, \ldots, v_m - v_1$ are linearly independent.*
(3) *The equations*

$$\lambda_1 v_1 + \cdots + \lambda_m v_m = 0$$
$$\lambda_1 + \cdots + \lambda_m = 0$$

*imply that $\lambda_1 = \cdots = \lambda_m = 0$.*
(4) *The vectors*

$$\begin{pmatrix} v_1 \\ 1 \end{pmatrix}, \ldots, \begin{pmatrix} v_m \\ 1 \end{pmatrix}$$

*are linearly independent in $\mathbb{R}^{d+1}$.*

## 2.2   Exercises and solutions

**Exercise 2.1.** Let $u, v \in \mathbb{R}^d$ with $u \neq v$. Prove that

$$L = \{(1-t)u + tv \mid t \in \mathbb{R}\}$$

is a line in $\mathbb{R}^d$ containing $u$ and $v$. Prove also that if $M$ is a line in $\mathbb{R}^d$ such that $u, v \in M$, then $M = L$.

**Solution 2.1.** By noting that

$$L = \{(1-t)u + tv \mid t \in \mathbb{R}\} = \{u + t(v-u) \mid t \in \mathbb{R}\},$$

where $v - u$ by assumption is non-zero, it follows that $L$ is a line containing $u$ and $v$.

If $M$ is a line that is, $M = \{x + t\alpha \mid t \in \mathbb{R}\}$ for some $x \in \mathbb{R}^d$ and $\alpha \in \mathbb{R}^d \setminus \{0\}$ such that $u, v \in M$, then we can choose $t_1, t_2 \in \mathbb{R}$, $t_1 \neq t_2$, so $u = x + t_1\alpha$ and $v = x + t_2\alpha$. It is now seen that an arbitrarily chosen element in $L$ is in $M$ since for all $t \in \mathbb{R}$ we have

$$\begin{aligned}
(1-t)u + tv &= (1-t)(x + t_1\alpha) + t(x + t_2\alpha) \\
&= x + ((1-t)t_1 + t \cdot t_2)\alpha.
\end{aligned} \tag{2.1}$$

On the other hand if $w \in M$, we have $w = x + s\alpha$ for some $s \in \mathbb{R}$, and then we just notice that we can find $t \in \mathbb{R}$ such that $s = (1-t)t_1 + t \cdot t_2$ since $t_1 \neq t_2$ and thus, $w \in L$ according to (2.1).

$$\star \quad \star \quad \star$$

**Exercise 2.2.** Let $u = (1,\ 1,\ 1)$ and $v = (1,\ 2\ ,3)$ be vectors in $\mathbb{R}^3$. Show that $u$ and $v$ are linearly independent and find $\alpha \in \mathbb{R}^3$ with

$$W = \{x \in \mathbb{R}^3 \mid \alpha^t x = 0\},$$

where $W$ is the subspace spanned by $u$ and $v$.

**Solution 2.2.** We have that $u$ and $v$ are linearly independent, since the homogeneous system of equations

$$c_1 \begin{pmatrix} 1 \\ 1 \\ 1 \end{pmatrix} + c_2 \begin{pmatrix} 1 \\ 2 \\ 3 \end{pmatrix} = \begin{pmatrix} 0 \\ 0 \\ 0 \end{pmatrix}$$

only can be solved with $c_1 = c_2 = 0$. If we use the first equation to get the relation $c_1 = -c_2$, we can substitute this into the second equation to obtain $c_2 = 0$. Having this, we find from the first equation that $c_1 = 0$.

To obtain $\alpha \in \mathbb{R}^3 \setminus \{0\}$ with $\alpha^t x = 0$ for all $x$ in the subspace spanned by $u$ and $v$ it is necessary and sufficient that we find a non-zero solution to the system $\alpha^t u = \alpha^t v = 0$. This would be the case for $\alpha = (1, -2, 1)$.

$$\star \quad \star \quad \star$$

**Exercise 2.3.** Let $W \subseteq \mathbb{R}^d$ be a subspace and suppose that $v_1, \ldots, v_r$ is a basis of $W$. Prove that

$$W' = \{u \in \mathbb{R}^d \mid u^t v_1 = \cdots = u^t v_r = 0\} \subseteq \mathbb{R}^d$$

is a subspace. Let $u_1, \ldots, u_s$ be a basis of $W'$ and $A$ the $s \times d$ matrix with these vectors as rows. Show that $s = d - r$ and

$$W = \{x \in \mathbb{R}^d \mid Ax = 0\}.$$

**Solution 2.3.** First, we will verify that $W'$ is a subspace. We note that $0 \in W'$. If $u, w \in W'$ and $\alpha \in \mathbb{R}$, then $(\alpha u + w)^t v_i = \alpha u^t v_i + w^t v_i = 0$ for $i = 1, \ldots, r$ and thus, $\alpha u + w \in W'$.

Now define the $d \times r$ matrix

$$B = \begin{pmatrix} -v_1^t- \\ \vdots \\ -v_r^t- \end{pmatrix} = \begin{pmatrix} B_1 & B_2 \end{pmatrix}$$

where $B_1$ is an $r \times r$ matrix and $B_2$ is an $r \times (d - r)$ matrix. By Corollary B.7, $B$ has $r$ linear independent columns and therefore, without loss of generality, we may assume that $B_1$ is invertible. Now note that

$$
\begin{aligned}
W' &= \{u \in \mathbb{R}^d \mid Bu = 0\} \\
&= \{u \in \mathbb{R}^d \mid \begin{pmatrix} I_r & B_1^{-1} B_2 \end{pmatrix} u = 0\} \\
&= \left\{ \begin{pmatrix} x \\ y \end{pmatrix} \in \mathbb{R}^d \,\middle|\, x \in \mathbb{R}^r,\ y \in \mathbb{R}^{d-r},\ x = Cy \right\}
\end{aligned}
$$

where $C = -B_1^{-1} B_2$. This shows that

$$\begin{pmatrix} c_1 \\ e_1 \end{pmatrix}, \ldots, \begin{pmatrix} c_{d-r} \\ e_{d-r} \end{pmatrix} \in W',$$

where $c_i$ is the $i$-th column of $C$ and $e_i$ is the $i$-th canonical basis vector in $\mathbb{R}^{d-r}$. Since these $d - r$ vectors are linear independent we conclude that $s \geq d - r$.

On the other hand, observe that if

$$y = a_1 v_1 + \cdots + a_r v_r = b_1 u_1 + \cdots + b_s u_s$$

then $y^t y = 0$. This gives linear independence of $\{v_1, \ldots, v_r, u_1, \ldots, u_s\}$ and consequently, $r + s \leq d$. This shows that $s = d - r$.

We will now show that $W = \{x \in \mathbb{R}^d | Ax = 0\}$. Since $r + s = d$, $\{v_1, \ldots, v_r, u_1, \ldots, u_s\}$ is a basis for $\mathbb{R}^d$. Therefore, we can write $x = w + w'$, where $w \in W$ and $w' \in W'$. Now we observe that $Ax = Aw'$ and $Aw' = 0$ if and only if $w' = 0$, from which the result follows.

$\star$ $\quad \star$ $\quad \star$

**Exercise 2.4.** Prove that $\{v \in \mathbb{R}^d \mid Av = b\}$ is an affine subspace of $\mathbb{R}^d$, where $A$ is an $m \times d$ matrix and $b \in \mathbb{R}^m$.

**Solution 2.4.** Let $u, v \in M = \{v \in \mathbb{R}^d \mid Av = b\}$ and $t \in \mathbb{R}$, and notice that

$$A((1 - t)u + tv) = (1 - t)Au + tAv = b,$$

which implies $(1 - t)u + tv \in M$.

$\star$ $\quad \star$ $\quad \star$

**Exercise 2.5.** Let $M$ be an affine subspace. Prove that $\{u - v \mid u, v \in M\}$ is a subspace.

**Solution 2.5.** It follows directly from the first part of the proof of Proposition 2.6.

$\star$ $\quad \star$ $\quad \star$

**Exercise 2.6.** Can you have two linearly independent vectors in $\mathbb{R}$? What about two affinely independent vectors?

**Solution 2.6.** We cannot have two linearly independent vectors in $\mathbb{R}$ which is an immediate consequence of Corollary B.3. Two vectors $u, v \in \mathbb{R}$ are affinely independent if and only if $(u, 1), (v, 1) \in \mathbb{R}^2$ are linearly independent according to Proposition 2.9(4), which is the case for $u \neq v$.

$\star$ $\quad \star$ $\quad \star$

**Exercise 2.7.** Decide if $(2, 1)$, $(3, 2)$ and $(5, 5)$ are on the same line in $\mathbb{R}^2$ applying Proposition 2.9.

**Solution 2.7.** If $(2, 1)$, $(3, 2)$ and $(5, 5)$ are on the same line in $\mathbb{R}^2$ it is the same as saying that they are affinely dependent. For instance, we can check this by using Proposition 2.9(2) and find that $(3, 2) - (2, 1) = (1, 1)$

and $(5,\ 5) - (2,\ 1) = (3,\ 4)$ are linearly independent and thus, the three points do not lie on the same line.

$$\star \quad \star \quad \star$$

**Exercise 2.8.** Let $S = \{(1,\ 1,\ 1), (2,\ 3,\ 4), (1,\ 2,\ 3), (2,\ 1,\ 0)\} \subseteq \mathbb{R}^3$. Compute the smallest affine subspace containing $S$.

**Solution 2.8.** Note that this exercise is reminiscent of Example 2.1. According to Proposition 2.5, aff$(S)$ is the smallest affine subspace containing $S$. It can be written as

$$\mathrm{aff}(S) = \left\{ \lambda_1 \begin{pmatrix} 1 \\ 1 \\ 1 \end{pmatrix} + \lambda_2 \begin{pmatrix} 2 \\ 3 \\ 4 \end{pmatrix} + \lambda_3 \begin{pmatrix} 1 \\ 2 \\ 3 \end{pmatrix} + \lambda_4 \begin{pmatrix} 2 \\ 1 \\ 0 \end{pmatrix} \,\Big|\, \sum_{i=1}^{4} \lambda_i = 1 \right\}.$$

It follows from Proposition 2.6 that we can write this affine subspace as a sum of a vector and a subspace. Using that $\lambda_1 = 1 - \lambda_2 - \lambda_3 - \lambda_4$ we find the representation

$$\mathrm{aff}(S) = \begin{pmatrix} 1 \\ 1 \\ 1 \end{pmatrix} + \left\{ t_1 \begin{pmatrix} 1 \\ 2 \\ 3 \end{pmatrix}) + t_2 \begin{pmatrix} 0 \\ 1 \\ 2 \end{pmatrix} + t_3 \begin{pmatrix} 1 \\ 0 \\ -1 \end{pmatrix} \,\Big|\, t_1, t_2, t_3 \in \mathbb{R} \right\}$$

$$= \begin{pmatrix} 1 \\ 1 \\ 1 \end{pmatrix} + \left\{ t_1 \begin{pmatrix} 1 \\ 2 \\ 3 \end{pmatrix} + t_2 \begin{pmatrix} 0 \\ 1 \\ 2 \end{pmatrix} \,\Big|\, t_1, t_2 \in \mathbb{R} \right\}.$$

This span can be represented using a vector that is normal to the subspace which amounts to finding a non-zero vector $\alpha \in \mathbb{R}^3$ that satisfies

$$\begin{pmatrix} 1 & 2 & 3 \\ 0 & 1 & 2 \end{pmatrix} \alpha = 0.$$

This could be $\alpha = (1,\ -2,\ 1)$ meaning we can write

$$\mathrm{aff}(S) = \begin{pmatrix} 1 \\ 1 \\ 1 \end{pmatrix} + \{(x, y, z) \in \mathbb{R}^3 \mid x - 2y + z = 0\}$$

$$= \{(x, y, z) \in \mathbb{R}^3 \mid x - 2y + z = 0\},$$

since $(1,\ 1,\ 1)\alpha = 0$.

$$\star \quad \star \quad \star$$

**Exercise 2.9.** Prove that $f(x) = h(x) + b$ is an affine map if $h : \mathbb{R}^d \to \mathbb{R}^n$ is a linear map and $b \in \mathbb{R}^n$. Prove that $h(x) = f(x) - f(0)$ is a linear map if $f : \mathbb{R}^d \to \mathbb{R}^n$ is an affine map.

**Solution 2.9.** Let $u, v \in \mathbb{R}^d$ and $t \in \mathbb{R}$. It follows that $f$ is an affine map, since

$$f((1-t)u + tv) = (1-t)(h(u) + b) + t(h(v) + b) = (1-t)f(u) + tf(v).$$

Now let $f$ be an affine map. It follows by induction that

$$f(\lambda_1 v_1 + \cdots + \lambda_m v_m) = \lambda_1 f(v_1) + \cdots + \lambda_m f(v_m)$$

where $\lambda_1 + \cdots + \lambda_m = 1$ and $v_1, \ldots, v_m \in \mathbb{R}^d$. Having this we note that

$$h(u + v) = f(u + v - 0) - f(0) = f(u) + f(v) - 2f(0) = h(u) + h(v)$$

for $u, v \in \mathbb{R}^d$. Furthermore, we have

$$\begin{aligned} h(\alpha u) &= f(\alpha u) - f(0) \\ &= f(\alpha u + (1-\alpha)0) - f(0) \\ &= \alpha f(u) + (1-\alpha)f(0) - \alpha f(0) - (1-\alpha)f(0) \\ &= \alpha h(x) \end{aligned}$$

when $\alpha \in \mathbb{R}$ and $u \in \mathbb{R}^d$.

$\star \quad \star \quad \star$

**Exercise 2.10.** Prove that you can have no more than $d + 1$ affinely independent vectors in $\mathbb{R}^d$.

**Solution 2.10.** Having more than $d+1$ affinely independent vectors in $\mathbb{R}^d$ corresponds to having more than $d$ linearly independent vectors in $\mathbb{R}^d$ by Proposition 2.9(2), which is a contradiction. Note, however, that you can have $d + 1$ affinely independent vectors in $\mathbb{R}^d$.

$\star \quad \star \quad \star$

**Exercise 2.11.** Let $v_0, \ldots, v_d$ be affinely independent points in $\mathbb{R}^d$. Prove that

$$f(x) = (\lambda_0, \lambda_1, \ldots, \lambda_d)$$

is a well defined affine map $f : \mathbb{R}^d \to \mathbb{R}^{d+1}$, where

$$x = \lambda_0 v_0 + \cdots + \lambda_d v_d$$

with $\lambda_0 + \cdots + \lambda_d = 1$.

**Solution 2.11.** The map is well defined since for any vector $x \in \mathbb{R}^d$, there is a unique $\lambda = (\lambda_0, \lambda_1, \ldots, \lambda_d) \in \mathbb{R}^{d+1}$ with $x = \lambda_0 v_0 + \lambda_1 v_1 + \cdots + \lambda_d v_d$ and $\lambda_0 + \lambda_1 + \cdots + \lambda_d = 1$. This follows from Proposition 2.9(4) which tells us that $(v_0, 1), (v_1, 1), \ldots, (v_d, 1) \in \mathbb{R}^{d+1}$ are linearly independent, hence the system of equations

$$\begin{pmatrix} v_0 & v_1 & \cdots & v_d \\ 1 & 1 & \cdots & 1 \end{pmatrix} \lambda = \begin{pmatrix} x \\ 1 \end{pmatrix}$$

has a unique solution. We now verify that the map is affine. Let $x, y \in \mathbb{R}^d$ and $t \in \mathbb{R}$, and note that if $f(x) = \lambda$ and $f(y) = \mu$, then $(1-t)x + ty = \sum_{i=0}^d ((1-t)\lambda_i + t\mu_i)v_i$ and $\sum_{i=0}^d ((1-t)\lambda_i + t\mu_i) = 1$. In conclusion,

$$
\begin{aligned}
f((1-t)x + ty) &= ((1-t)\lambda_0 + t\mu_0, \ldots, (1-t)\lambda_d + t\mu_d) \\
&= (1-t)f(x) + tf(y).
\end{aligned}
$$

$$\star \quad \star \quad \star$$

**Exercise 2.12.** Prove that a non-empty open subset $U \subseteq \mathbb{R}^d$ has dimension $\dim U = d$. Show that a subset $S \subseteq \mathbb{R}^d$ with $\dim S = d$ contains a non-empty open subset.

**Solution 2.12.** Since $U$ is non-empty and open we can choose $x \in U$ and find $\epsilon > 0$ such that $B(x, \epsilon) \subseteq U$ (see §A.4). Let $v_1, \ldots, v_d \in \mathbb{R}^d$ be linearly independent vectors with $|v_i| \leq \epsilon$ for $i = 1, \ldots, d$. Then $x + v_i \in U$ for $i = 1, \ldots, d$, and if we let

$$V = \{x, x + v_1, \ldots, x + v_d\} \subseteq U,$$

it follows from Proposition 2.5 that $\mathrm{aff}(V) \subseteq \mathrm{aff}(U)$ which shows $\dim \mathrm{aff}(V) \leq \dim \mathrm{aff}(U)$. Proposition 2.9 gives that $\mathrm{aff}(V) = x + W$ where $W = \{t_1 v_1 + \cdots + t_d v_d \mid t_1, \ldots, t_d \in \mathbb{R}\}$ and therefore we have that

$$\dim U = \dim \mathrm{aff}(U) \geq \dim \mathrm{aff}(V) = \dim W = d.$$

On the other hand $U \subseteq \mathbb{R}^d$, which gives $\dim U \leq d$ and hence, we conclude that $\dim U = d$.

The latter part of the exercise is incorrectly stated: a counterexample of a set not containing any open subset, but where the dimension is two, is

$$S = \{(0,0), (1,0), (0,1)\} \subseteq \mathbb{R}^2,$$

since $\mathrm{aff}(S) = \mathbb{R}^2$.

However, the result will indeed hold if $S$ is assumed to be a convex subset (cf. Definition 3.1). To prove this statement we need to make use of some results from later chapters and therefore, one may find it useful to return to this part after reading (at least) Chapter 3. If $\dim S = d$ we may find affinely independent vectors $v_0, v_1, \ldots, v_d \in S$ and in addition, if $S$ is convex it follows from Proposition 3.4 that $C = \text{conv}(\{v_0, v_1, \ldots, v_d\}) \subseteq S$. In Exercise 6.6 it is shown that

$$\tfrac{1}{d+1} (v_0 + v_1 + \cdots + v_d) \in \text{int}(C) \subseteq S.$$

As $\text{int}(C)$ is open by definition the result is proved.

# Chapter 3

# Convex subsets

## 3.1 Introduction

The simplicity of the definition of a convex set stands in great contrast to its ramifications. When written concisely it takes up less than a line, yet it will be the foundation for much of the remaining material. Many of the exercises have convexity at the heart of the solution, and there is no doubt that solving a fair amount of them will give familiarity with the definition of convexity. Concepts such as convex hulls and (convex) cones will also be important in the coming chapters, and solving exercises related to these is therefore beneficial.

**Definition 3.1.** A subset $C \subseteq \mathbb{R}^d$ is called *convex* if it contains the line segment between any two of its points:

$$(1 - t)u + tv \in C$$

for every $u, v \in C$ and every $t \in \mathbb{R}$ with $0 \leq t \leq 1$.

**Definition 3.3.** The *convex hull of a subset* $S \subseteq \mathbb{R}^d$ is the set of all convex linear combinations of elements from $S$ i.e.,

$$\text{conv}(S) := \{\lambda_1 v_1 + \cdots + \lambda_m v_m \mid m \geq 1,$$
$$v_1, \ldots, v_m \in S, \lambda_1, \ldots, \lambda_m \geq 0 \text{ and } \lambda_1 + \cdots + \lambda_m = 1\}.$$

If $S$ is a finite subset, $\text{conv}(S)$ is called a *polytope*. If $S \subseteq \mathbb{R}^2$ is a finite subset, we call $\text{conv}(S)$ a (convex) *polygon*.

Besides the convex hulls, another important family of convex sets are the convex cones. A set $K \subseteq \mathbb{R}^n$ is a cone if $x \in K$ implies $\lambda x \in K$ for every $\lambda \geq 0$. A cone is a convex cone if it is convex as well. A particular convex cone (see Exercise 3.18) is the recession cone of a convex set $C \subseteq \mathbb{R}^n$, and it is defined as

$$\mathrm{rec}(C) = \{d \in \mathbb{R}^n \mid x + d \in C \text{ for every } x \in C\}.$$

**Proposition 3.4.** *The convex hull,* $\mathrm{conv}(S)$, *of a subset* $S \subseteq \mathbb{R}^d$ *is a convex subset. It is the smallest convex subset containing* $S$.

Another way of formulating the last part of Proposition 3.4 is that if $S \subseteq C$ and $C$ is a convex set, $\mathrm{conv}(S) \subseteq C$.

**Definition 3.7.** Let $C \subseteq \mathbb{R}^d$ be a convex subset. A subset $F \subseteq C$ is called a *face* of $C$ if $F$ is convex and for every $x, y \in C$ and $0 < \lambda < 1$,

$$(1 - \lambda)x + \lambda y \in F$$

implies that $x, y \in F$.

A face $F \subseteq C$ of a convex set $C \subseteq \mathbb{R}^n$ is called *exposed* if it is of the form

$$F = \{z \in C \mid \alpha^t z \leq \alpha^t x, \text{ for every } x \in C\} \qquad (3.1)$$

for some $\alpha \in \mathbb{R}^n$.

**Corollary 3.15.** *Let* $V = \{v_1, \ldots, v_m\} \subseteq \mathbb{R}^d$. *If* $v \in \mathrm{conv}(V)$, *then* $v$ *belongs to the convex hull of an affinely independent subset of* $V$.

## 3.2   Exercises and solutions

**Exercise 3.1.** Let $P = \{x \in \mathbb{R}^d \mid Ax \leq b\}$ be a polyhedron in $\mathbb{R}^d$. Prove that $(1 - t)x + ty \in P$ if $x, y \in P$ and $0 \leq t \leq 1$.

**Solution 3.1.** Given $x, y \in P$ we see that

$$A((1 - t)x + ty) = (1 - t)Ax + tAy \leq (1 - t)b + tb = b$$

for $0 \leq t \leq 1$ and hence, $(1 - t)x + ty \in P$.

$$\star \quad \star \quad \star$$

**Exercise 3.2.** Let $A, B \subseteq \mathbb{R}^d$ be convex subsets. Prove that $A \cap B$ is a convex subset. Give an example showing that $A \cup B$ does not have to be a convex subset.

**Solution 3.2.** Suppose that $u, v \in A \cap B$. Since $A$ is a convex set, $(1 - t)u + tv \in A$ for every $t \in [0, 1]$. The same applies for $B$, and we conclude that $(1 - t)u + tv \in A \cap B$ for every $t \in [0, 1]$.

An example where $A \cup B$ is not a convex subset could be $A, B \subseteq \mathbb{R}$ given by $A = [0, 1]$ and $B = [2, 3]$. Choosing, for instance, $x = 1, y = 2$, and $\lambda = 1/2$, it follows that $x, y \in A \cup B$ but $(1 - \lambda)x + \lambda y \notin A \cup B$.

$$\star \quad \star \quad \star$$

**Exercise 3.3.** Prove in detail that an affine half space is a convex subset and that a polyhedron

$$P = \{x \in \mathbb{R}^d \mid Ax \leq b\}$$

is a convex subset of $\mathbb{R}^d$.

**Solution 3.3.** Consider the affine half space $H^- = \{x \in \mathbb{R}^d \mid \alpha^t x \leq \beta\}$ for $\alpha \in \mathbb{R}^d$ and $\beta \in \mathbb{R}$, and let $x, y \in H^-$ and $\lambda \in [0, 1]$. It is seen that

$$\alpha^t((1 - \lambda)x + \lambda y) = (1 - \lambda)\alpha^t x + \lambda \alpha^t y \leq (1 - \lambda)\beta + \lambda \beta = \beta$$

which implies that $(1 - \lambda)x + \lambda y \in H^-$, hence $H^-$ is convex. Furthermore, since $\alpha$ and $\beta$ were arbitrary, this also shows that $H^+$ is convex.

We can recognize that $P$ is a finite intersection of half spaces which again is a convex set by Exercise 3.2. Alternatively, Exercise 3.1 gives us that $P = \{x \in \mathbb{R}^d \mid Ax \leq b\}$ is convex.

$$\star \quad \star \quad \star$$

**Exercise 3.4.** Let $A$ be a convex subset of $\mathbb{R}^d$. Prove that

$$A + z := \{x + z \mid x \in A\}$$

and

$$\lambda A := \{\lambda x \mid x \in A\}$$

are convex subsets of $\mathbb{R}^d$ for $z \in \mathbb{R}^d$ and $\lambda \in \mathbb{R}$. Let $B$ be a convex subset of $\mathbb{R}^d$. Prove that

$$A + B := \{x + y \mid x \in A, \, y \in B\}$$

is a convex subset of $\mathbb{R}^d$.

**Solution 3.4.** For $x, y \in A$ and $t \in [0, 1]$ we find

$$(1 - t)(x + z) + t(y + z) = (1 - t)x + ty + z.$$

Notice that $(1 - t)x + ty \in A$ since $A$ is convex and thus, by definition of $A + z$, $(1 - t)x + ty + z \in A + z$, showing that $A + z$ is a convex subset of $\mathbb{R}^d$.

Similarly, it holds that

$$(1 - t)\lambda x + t\lambda y = \lambda((1 - t)x + ty).$$

By assumption $(1 - t)x + ty \in A$ and hence, $\lambda((1 - t)x + ty) \in \lambda A$.

For $x_1, x_2 \in A$ and $y_1, y_2 \in B$,

$$(1 - t)(x_1 + y_1) + t(x_2 + y_2) = (1 - t)x_1 + tx_2 + (1 - t)y_1 + ty_2$$

for $t \in [0, 1]$. Since $(1 - t)x_1 + tx_2 \in A$ and $(1 - t)y_1 + ty_2 \in B$, the equation above shows the convexity of $A + B$.

$$\star \quad \star \quad \star$$

**Exercise 3.5.** Let $v_1, v_2, v_3 \in \mathbb{R}^n$. Show that

$$\{(1 - \lambda)v_3 + \lambda((1 - \mu)v_1 + \mu v_2) \mid \lambda \in [0, 1], \mu \in [0, 1]\}$$
$$= \{\lambda_1 v_1 + \lambda_2 v_2 + \lambda_3 v_3 \mid \lambda_1, \lambda_2, \lambda_3 \geq 0, \ \lambda_1 + \lambda_2 + \lambda_3 = 1\}.$$

**Solution 3.5.** Set

$$A = \{(1 - \lambda)v_3 + \lambda((1 - \mu)v_1 + \mu v_2) \mid \lambda \in [0, 1], \ \mu \in [0, 1]\}$$

and

$$B = \{\lambda_1 v_1 + \lambda_2 v_2 + \lambda_3 v_3 \mid \lambda_1, \lambda_2, \lambda_3 \geq 0, \ \lambda_1 + \lambda_2 + \lambda_3 = 1\}.$$

If $v \in A$, $v = (1 - \lambda)v_3 + \lambda((1 - \mu)v_1 + \mu v_2)$ for some $\lambda, \mu \in [0, 1]$. Now let $\lambda_1 = \lambda(1 - \mu)$, $\lambda_2 = \lambda\mu$, and $\lambda_3 = 1 - \lambda$. After verifying that $\lambda_1, \lambda_2, \lambda_3 \geq 0$ and $\lambda_1 + \lambda_2 + \lambda_3 = 1$ under the assumptions of $\lambda$ and $\mu$, we conclude that $v \in B$. If on the other hand $v \in B$, we can write $v = \lambda_1 v_1 + \lambda_2 v_2 + \lambda_3 v_3$ for some $\lambda_1, \lambda_2$ and $\lambda_3$ that satisfy the given conditions. By solving the system

$$\lambda\mu \quad = \lambda_2$$
$$1 - \lambda \quad = \lambda_3$$

we obtain $\lambda = 1 - \lambda_3$, $\mu = \lambda_2/(1 - \lambda_3) = (1 - \lambda_1 - \lambda_3)/(1 - \lambda_3)$ provided that $\lambda_3 \neq 1$ (notice that $\lambda(1 - \mu) = \lambda_1$ is automatically satisfied). Since $\lambda_1, \lambda_2, \lambda_3 \geq 0$ and $\lambda_1 + \lambda_2 + \lambda_3 = 1$ we have that $\lambda, \mu \in [0, 1]$ and thus, $v \in A$. If $\lambda_3 = 1$ then $v = v_3 \in A$.

This exercise gives a geometric interpretation of the convex hull of three vectors. More specifically, a point in $\mathrm{conv}(\{v_1, v_2, v_3\})$ must lie on the line segment connecting $v_3$ and some point on the line segment between $v_1$ and $v_2$ (see Figure 3.1).

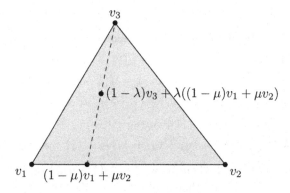

**Figure 3.1:** Illustration of points in $\mathrm{conv}(\{v_1, v_2, v_3\})$.

$$\star \quad \star \quad \star$$

**Exercise 3.6.** Sketch the convex hull of

$$S = \{(0,0), (1,0), (1,1)\} \subseteq \mathbb{R}^2.$$

Write $\mathrm{conv}(S)$ as the intersection of three half planes.

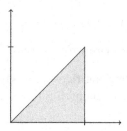

**Figure 3.2:** The convex hull $\mathrm{conv}(S)$.

**Solution 3.6.** In Figure 3.2 conv$(S)$ is sketched.

From the illustration it appears that we can write conv$(S)$ as

$$\text{conv}(S) = \{(x, y) \in \mathbb{R}^2 \mid -y \leq 0, \ x \leq 1, \ y - x \leq 0\}.$$

$$\star \quad \star \quad \star$$

**Exercise 3.7.** Let $u_1, u_2, v_1, v_2 \in \mathbb{R}^n$. Show that

$$\text{conv}(\{u_1, u_2\}) + \text{conv}(\{v_1, v_2\}) = \text{conv}(\{u_1 + v_1, u_1 + v_2, u_2 + v_1, u_2 + v_2\}).$$

**Solution 3.7.** Let $x \in \text{conv}(\{u_1, u_2\}) + \text{conv}(\{v_1, v_2\})$ and write

$$\begin{aligned}
x &= (1 - \lambda)u_1 + \lambda u_2 + (1 - \mu)v_1 + \mu v_2 \\
&= (1 - \mu)((1 - \lambda)u_1 + \lambda u_2) + \mu((1 - \lambda)u_1 + \lambda u_2) \\
&\quad + (1 - \lambda)((1 - \mu)v_1 + \mu v_2) + \lambda((1 - \mu)v_1 + \mu v_2) \\
&= (1 - \lambda)(1 - \mu)(u_1 + v_1) + (1 - \lambda)\mu(u_1 + v_2) \\
&\quad + \lambda(1 - \mu)(u_2 + v_1) + \lambda\mu(u_2 + v_2) \\
&= \lambda_1(u_1 + v_1) + \lambda_2(u_1 + v_2) + \lambda_3(u_2 + v_1) + \lambda_4(u_2 + v_2)
\end{aligned}$$

for some $\lambda, \mu \in [0, 1]$, and where $\lambda_1 = (1 - \lambda)(1 - \mu)$, $\lambda_2 = (1 - \lambda)\mu$, $\lambda_3 = \lambda(1 - \mu)$, and $\lambda_4 = \lambda\mu$. Note that $\lambda, \mu \in [0, 1]$ implies that $\lambda_1, \lambda_2, \lambda_3, \lambda_4 \geq 0$ and $\lambda_1 + \lambda_2 + \lambda_3 + \lambda_4 = 1$, hence $x \in \text{conv}(\{u_1 + v_1, u_1 + v_2, u_2 + v_1, u_2 + v_2\})$.

Suppose now that $x \in \text{conv}(\{u_1 + v_1, u_1 + v_2, u_2 + v_1, u_2 + v_2\})$ and write

$$\begin{aligned}
x &= \lambda_1(u_1 + v_1) + \lambda_2(u_1 + v_2) + \lambda_3(u_2 + v_1) + \lambda_4(u_2 + v_2) \\
&= \big[(\lambda_1 + \lambda_2)u_1 + (\lambda_3 + \lambda_4)u_2\big] + \big[(\lambda_1 + \lambda_3)v_1 + (\lambda_2 + \lambda_4)v_2\big] \quad (3.2)
\end{aligned}$$

for suitable $\lambda_1, \lambda_2, \lambda_3, \lambda_4 \geq 0$ with $\lambda_1 + \lambda_2 + \lambda_3 + \lambda_4 = 1$. The representation (3.2) shows that $x \in \text{conv}(\{u_1, u_2\}) + \text{conv}(\{v_1, v_2\})$.

$$\star \quad \star \quad \star$$

**Exercise 3.8.** Let $S \subseteq \mathbb{R}^n$ be a convex subset and $v \in \mathbb{R}^n$. Show that

$$\{(1 - \lambda)s + \lambda v \mid \lambda \in [0, 1], s \in S\}$$

is a convex subset. Hint: compare with Exercise 3.5.

**Solution 3.8.** Instead of showing that $\{(1 - \lambda)s + \lambda v \mid \lambda \in [0, 1], \ s \in S\}$ is a convex subset it is (when comparing to Exercise 3.5) tempting to try to show that

$$\{(1 - \lambda)s + \lambda v \mid \lambda \in [0, 1], \ s \in S\} = \text{conv}(S \cup \{v\}),$$

since it will imply that the subset is convex by Proposition 3.4.

Suppose that we can write $u = (1 - \lambda)s + \lambda v$ for some $\lambda \in [0,1]$ and $s \in S$. Since $s, v \in \text{conv}(S \cup \{v\})$ we have that $u \in \text{conv}(S \cup \{v\})$. If on the other hand $u \in \text{conv}(S \cup \{v\})$, we may find $s_1, \ldots, s_m \in S$ for $m \geq 1$ such that $u = \lambda_0 v + \lambda_1 s_1 + \cdots + \lambda_m s_m$ for some $\lambda_0, \ldots, \lambda_m \geq 0$ with $\sum_{i=0}^m \lambda_i = 1$. We may assume that $u \neq v$, equivalently $\lambda_0 \neq 1$, and in this case write

$$u = \lambda_0 v + (1 - \lambda_0) \left( \frac{\lambda_1}{1 - \lambda_0} s_1 + \cdots + \frac{\lambda_m}{1 - \lambda_0} s_m \right). \tag{3.3}$$

Observe that

$$\frac{\lambda_1}{1 - \lambda_0} s_1 + \cdots + \frac{\lambda_m}{1 - \lambda_0} s_m \in \text{conv}(\{s_1, \ldots, s_m\}) \subseteq S$$

and hence, $u \in \{(1 - \lambda)s + \lambda v \mid \lambda \in [0,1], \ s \in S\}$ by (3.3).

$$\star \quad \star \quad \star$$

**Exercise 3.9.** Let $C = \text{conv}(\{v_1, \ldots, v_m\})$ and $P = C^\circ$ for $v_1, \ldots, v_m \in \mathbb{R}^d$. Prove that

$$P = \{\alpha \in \mathbb{R}^d \mid \alpha^t v_1 \leq 1, \ldots, \alpha^t v_m \leq 1\}$$

and that $P$ is a polyhedron. Compute and sketch for $C$ given in Example 3.6. Prove in general that $P$ is bounded if 0 is an interior point of $C$.

**Solution 3.9.** Let $\alpha \in \mathbb{R}^d$ satisfy that $\alpha^t v_i \leq 1$ for $i = 1, \ldots m$. Then

$$\alpha^t(\lambda_1 v_1 + \cdots + \lambda_m v_m) = \lambda_1 \alpha^t v_1 + \cdots + \lambda_m \alpha^t v_m \leq \lambda_1 + \cdots + \lambda_m = 1$$

for any $\lambda_1, \ldots, \lambda_m \geq 0$ with $\lambda_1 + \cdots + \lambda_m = 1$, and we conclude that $\alpha \in P$. By definition we have $P \subseteq \{\alpha \in \mathbb{R}^d \mid \alpha^t v_1 \leq 1, \ldots, \alpha^t v_m \leq 1\}$ which gives the desired conclusion.

Now we may recognize $P$ as a polyhedron, since it is given as the set of solutions to finitely many, more precisely $m$, linear inequalities.

In Example 3.6 we have $C = \text{conv}(\{(0,1), (3,2), (4,0)\})$ and hence,

$$P = \{(\alpha_1, \alpha_2) \in \mathbb{R}^2 \mid \alpha_2 \leq 1, \ 3\alpha_1 + 2\alpha_2 \leq 1, \ 4\alpha_1 \leq 1\}.$$

This unbounded set is depicted in Figure 3.3. If, on the other hand, $0 \in \text{int}(C)$ then $P$ will be bounded. Assume for contradiction that $P$ is unbounded. Then there exists $\epsilon > 0$ such that $B(0, \epsilon) = \{y \in \mathbb{R}^d \mid |y| \leq \epsilon\} \subseteq C$

(see §A.4-5). Choose $\alpha = (\alpha_1, \ldots, \alpha_d) \in P$ with $|\alpha_i| > \epsilon^{-1}$ for some $i$. Without loss of generality we assume that $|\alpha_i| = \alpha_i$. Let $e_i$ be the $i$-th canonical basis vector in $\mathbb{R}^d$ and consider the vector $\epsilon e_i \in B(0, \epsilon) \subseteq C$. Note that

$$\alpha^t \epsilon e_i = \alpha_i \epsilon > \epsilon^{-1} \epsilon = 1.$$

Consequently $\alpha \notin P$, which is a contradiction, and we conclude that $P$ is bounded.

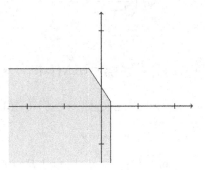

**Figure 3.3:** The polar of $C$ from Exercise 3.9.

⋆   ⋆   ⋆

**Exercise 3.10.** If $F \subseteq G \subseteq C$ are convex subsets of $\mathbb{R}^d$, prove that $F$ is a face of $C$ if $F$ is a face of $G$ and $G$ is a face of $C$.

**Solution 3.10.** Let $x, y \in C$ and $0 < \lambda < 1$ such that $(1 - \lambda)x + \lambda y \in F$. Since $F \subseteq G$ and $G$ is a face of $C$, $x, y \in G$. Additionally, $F$ is a face of $G$ implying that $x, y \in F$.

⋆   ⋆   ⋆

**Exercise 3.11.** Give an example of a convex subset $C \subseteq \mathbb{R}^d$ and a face $F \subseteq \mathbb{R}^d$, which is not exposed (hint: think about stretching a disc).

**Solution 3.11.** Consider the convex subset of $\mathbb{R}^2$ given by

$$C = \{(x, y) \in \mathbb{R}^2 \mid (x - 1)^2 + y^2 \leq 1\} \cup ([0, 1] \times [-1, 1])$$

(see Figure 3.4). The point $v = (1, 1) \in C$ is a zero-dimensional face of $C$ (an extreme point) but we cannot find $\alpha \in \mathbb{R}^2$ such that $F = \{v\}$ in the notation of (3.1). We verify that $v$ is an extreme point by first noting that if $(x, y) \in C$ then $y \leq 1$. Therefore, if two points $(x_1, y_1), (x_2, y_2) \in C$ satisfy that $v = (1 - \lambda)(x_1, y_1) + \lambda(x_2, y_2)$ for some $\lambda \in (0, 1)$, then $y_1 = y_2 = 1$. If these points are distinct we may assume that $x_2 > 1$, but this implies $(x_2, 1) \notin C$. Thus, we must have that $(x_1, y_1) = (x_2, y_2) = v$ and $v$ is an extreme point of $C$. Finally, we need to argue that we cannot find $\alpha = (\alpha_1, \alpha_2) \in \mathbb{R}^2$ with $\alpha^t v < \alpha^t u$ for every $u \in C \setminus \{v\}$. By comparing with $u_1 = (0, 1)$ and $u_2 = (1, 0)$, it must be the case that $\alpha_1, \alpha_2 < 0$. First, one may verify that

$$w_s = \left(\sqrt{1 - s^2} + 1, s\right) \in \{(x, y) \in \mathbb{R}^2 \mid (x - 1)^2 + y^2 = 1\} \subseteq C$$

if $s \in [0, 1)$. It suffices to show that $s$ can be chosen such that

$$\alpha^t w_s = \alpha_1 \sqrt{1 - s^2} + \alpha_1 + \alpha_2 s \leq \alpha_1 + \alpha_2 = \alpha^t v$$

or equivalently,

$$\sqrt{\frac{1 + s}{1 - s}} \geq \frac{\alpha_2}{\alpha_1}.$$

Since the left-hand side approaches infinity as $s \uparrow 1$ and $\alpha_2/\alpha_1$ is a fixed number, we conclude that there exists $s \in [0, 1)$ such that $\alpha^t w_s \leq \alpha^t v$.

The seemingly obscure construction of $w_s$ actually has a very intuitive geometric interpretation. As $s$ goes from 0 to 1, $w_s$ moves from $(2, 0)$ to $v$ along the circle arc. When considering $\alpha_1, \alpha_2 < 0$, $\{u \in \mathbb{R}^2 \mid \alpha^t u = \alpha^t v\}$ is a (strictly) downward sloping line containing $v$. Any point $u$ above the line has $\alpha^t u \leq \alpha^t v$, and this will especially be the case for a circle arc segment starting from $v$ (see Figure 3.4).

$\star \quad \star \quad \star$

**Exercise 3.12.** Prove that $C \setminus F$ is a convex subset if $F$ is a face of a convex subset $C$. Is it true that $F \subseteq C$ is a face if $C \setminus F$ is a convex subset?

**Solution 3.12.** Let $x, y \in C \setminus F$ and $0 < \lambda < 1$. Since $C$ is convex it must be that $(1 - \lambda)x + \lambda y \in C \setminus F$ or $(1 - \lambda)x + \lambda y \in F$. We cannot have $(1 - \lambda)x + \lambda y \in F$ since this would imply $x, y \in F$.

For instance, the converse statement fails if $C = [0, 1]$ and $F = [0, 1/2]$ (consider $0, 1 \in C$ and $\lambda = 1/2$).

$\star \quad \star \quad \star$

**Exercise 3.13.** Let $X = \{x_1, \ldots, x_n\} \subseteq \mathbb{R}^d$.

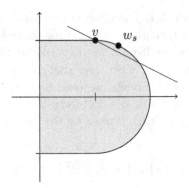

**Figure 3.4:** The set $C$ where $v = (1,1)$ is an example of a face that is *not* exposed.

(i) Prove that if $z \in \text{conv}(X)$ is an extreme point, then $z \in X$.

(ii) Suppose that $z \notin X$. Prove that $z$ is an extreme point of $\text{conv}(\{z\} \cup X)$ if and only if

$$z \notin \text{conv}(X).$$

This means that the extreme points in a convex hull consists of the "non-redundant generators" (compare this with [U.C., Figure 3.6]).

**Solution 3.13.**     (i) If $n = 1$ then $z = x_1$. Now let $n \geq 2$ and suppose that the statement holds for $n - 1$ and that $z \in \text{conv}(X)$ is an extreme point. Then we may find $\lambda_1, \ldots, \lambda_n \geq 0$ with $\lambda_1 + \cdots + \lambda_n = 1$ such that

$$z = \lambda_1 x_1 + \cdots + \lambda_n x_n.$$

If $z \neq x_1$ we have that $\lambda_1 < 1$ and we can write

$$z = \lambda_1 x_1 + (1 - \lambda_1) \left[ \frac{\lambda_2}{1 - \lambda_1} x_2 + \cdots + \frac{\lambda_n}{1 - \lambda_1} x_n \right].$$

Since $z$ is an extreme point, $z = x_1$ or

$$z = \frac{\lambda_2}{1 - \lambda_1} x_2 + \cdots + \frac{\lambda_n}{1 - \lambda_1} x_n,$$

and as the former does not hold, $z \in \text{conv}(\{x_2, \ldots, x_n\})$ which by our induction hypothesis implies that $z \in \{x_2, \ldots, x_n\}$.

(ii) We will show that if $z \in \text{conv}(X)$ then $z$ cannot be an extreme point of $\text{conv}(\{z\} \cup X)$. If $z \in \text{conv}(X)$, $z$ is not an extreme point of $\text{conv}(X)$ as this would imply that $z \in X$ by (i). By the inclusion $\text{conv}(X) \subseteq \text{conv}(\{z\} \cup X)$, $z$ is not an extreme point of $\text{conv}(\{z\} \cup X)$.

Now assume that $z \notin \text{conv}(X)$ and note that choosing $\lambda_1, \ldots, \lambda_{n+1} \geq 0$ with $\lambda_1 + \cdots + \lambda_{n+1} = 1$ and

$$z = \lambda_{n+1} z + \sum_{i=1}^{n} \lambda_i x_i \tag{3.4}$$

gives

$$z = \sum_{i=1}^{n} \frac{\lambda_i}{1 - \lambda_{n+1}} x_i$$

if $\lambda_{n+1} < 1$. This implies that $z \in \text{conv}(X)$ resulting in a contradiction, hence we have the representation (3.4) only if $\lambda_{n+1} = 1$. Now let $x, y \in \text{conv}(\{z\} \cup X)$. If $z = (1 - \lambda)x + \lambda y$ for $0 \leq \lambda \leq 1$ then

$$z = (1 - \lambda) \left( \lambda_{n+1}^x z + \sum_{i=1}^{n} \lambda_i^x x_i \right) + \lambda \left( \lambda_{n+1}^y z + \sum_{i=1}^{n} \lambda_i^y x_i \right)$$

$$= [(1 - \lambda)\lambda_{n+1}^x + \lambda \lambda_{n+1}^y]z + \sum_{i=1}^{n} [(1 - \lambda)\lambda_i^x + \lambda \lambda_i^y]x_1$$

for some $\lambda_i^x, \lambda_i^y \geq 0$, $i = 1, \ldots, n+1$, with $\sum_{i=1}^{n+1} \lambda_i^x = \sum_{i=1}^{n+1} \lambda_i^y = 1$. We have previously argued that the above representation implies that $(1 - \lambda)\lambda_{n+1}^x + \lambda \lambda_{n+1}^y = 1$, and this can only happen if either $\lambda_{n+1}^x = 1$ or $\lambda_{n+1}^y = 1$. In other words $z = x$ or $z = y$, showing that $z$ is an extreme point.

In Example 3.6, it is argued that $(2, 1) \in \text{conv}(\{(0, 1), (3, 2), (4, 0)\})$, and in Figure 3.7 it is seen that $(2, 1)$ is not an extreme point of the convex hull of $(0, 1), (3, 2), (4, 0)$, and $(2, 1)$. This can be seen as a general consequence of (ii) with $z = (2, 1)$. In addition, one may agree that $(0, 1)$, $(3, 2)$, and $(4, 0)$ seem to be extreme points of the convex hull, which is in line with (i) that tells us that the points are in the generating set.

**Exercise 3.14.** Prove in detail that

$$C = \{(x_1, x_2) \in \mathbb{R}^2 \mid x_1^2 + x_2^2 \leq 1\}$$

is a convex subset of $\mathbb{R}^2$. What are the extreme points of $C$? Can you prove it?

**Solution 3.14.** Observe that we can write $C$ as

$$C = \left\{ \begin{pmatrix} x_1 \\ x_2 \end{pmatrix} \in \mathbb{R}^2 \;\middle|\; \left| \begin{pmatrix} x_1 \\ x_2 \end{pmatrix} \right| \leq 1 \right\}.$$

Therefore, by the triangle inequality (Theorem A.2), $C$ is a convex subset.

We show that $\text{ext}(C) = \{x \in \mathbb{R}^2 \mid |x| = 1\}$. Let $x = (1 - \lambda)u + \lambda v$ for some $0 < \lambda < 1$ and $u, v \in C$ such that $|x| = 1$. Since $|u|, |v| \leq 1$ and $1 \leq (1 - \lambda)|u| + \lambda|v|$ by the triangle inequality, we have $|u| = |v| = 1$. Consequently, $|(1-\lambda)u + \lambda v| = |(1-\lambda)u| + |\lambda v|$ and it follows from Lemma A.1 and Theorem A.2 that there exists $\mu > 0$ such that $(1 - \lambda)u = \mu\lambda v$. From this we conclude that $u = v = x$.

Next we show that all extreme points are given this way. Let $x \in C$ satisfy $0 < |x| < 1$. From the identity

$$x = |x|\frac{x}{|x|} + (1 - |x|)0$$

it is evident that $x$ is not an extreme point. The point $0 \in \mathbb{R}^2$ is not extreme either, since we can write this element as a convex linear combination of two other points in $C$.

$$\star \quad \star \quad \star$$

**Exercise 3.15.** Recall the notation

$$A + B = \{u + v \mid u \in A, v \in B\} \subseteq \mathbb{R}^n$$

for two subsets $A, B \subseteq \mathbb{R}^n$ and let $[u, v] := \text{conv}(\{u, v\})$ for $u, v \in \mathbb{R}^n$.

(i) Show that
$$[u, v] + \{w\} = [u + w, v + w]$$
    for $u, v, w \in \mathbb{R}^n$.

(ii) Sketch
$$P = \left[ \begin{pmatrix} 1 \\ 1 \end{pmatrix}, \begin{pmatrix} 1 \\ 2 \end{pmatrix} \right] + \left[ \begin{pmatrix} 2 \\ 1 \end{pmatrix}, \begin{pmatrix} 3 \\ 2 \end{pmatrix} \right]$$
    along with its extreme points in the plane.

(iii) Let
$$Q = P + \left[ \begin{pmatrix} 3 \\ 1 \end{pmatrix}, \begin{pmatrix} 4 \\ 1 \end{pmatrix} \right].$$

Write $Q$ as a convex hull of the minimal number of points and as an intersection of half planes.

(iv) Let $A$ and $B$ be convex sets and $u_0 \in A$ a point in $A$, which is *not* extreme. Show that $u_0 + b \in A + B$ is not extreme in $A + B$ for any $b \in B$.

(v) Show that
$$T = [x, y] + [z, w]$$
has at most 4 extreme points for $x, y, z, w \in \mathbb{R}^n$. Can $T$ have 3 extreme points? 2?

(vi) Let $L_i = [u_i, v_i]$ for $i = 1, \ldots, m$, where $u_i, v_i \in \mathbb{R}^n$. Give an upper bound for how many extreme points
$$Z = L_1 + \cdots + L_m \tag{3.5}$$
can have. Show that $Z$ is the image of the unit cube $[0, 1]^m \subseteq \mathbb{R}^m$ under a suitable affine map.

The Minkowski sum of finitely many line segments (as in (3.5)) is called a *zonotope*.

**Solution 3.15.** (i) This follows from the identity $(1 - \lambda)u + \lambda v + w = (1 - \lambda)(u + w) + \lambda(v + w)$ for $0 \le \lambda \le 1$.

(ii) According to Exercise 3.7,
$$P = \left[ \begin{pmatrix} 1 \\ 1 \end{pmatrix}, \begin{pmatrix} 1 \\ 2 \end{pmatrix} \right] + \left[ \begin{pmatrix} 2 \\ 1 \end{pmatrix}, \begin{pmatrix} 3 \\ 2 \end{pmatrix} \right]$$
$$= \text{conv}\left( \left\{ \begin{pmatrix} 1 \\ 1 \end{pmatrix} + \begin{pmatrix} 2 \\ 1 \end{pmatrix}, \begin{pmatrix} 1 \\ 1 \end{pmatrix} + \begin{pmatrix} 3 \\ 2 \end{pmatrix}, \begin{pmatrix} 1 \\ 2 \end{pmatrix} + \begin{pmatrix} 2 \\ 1 \end{pmatrix}, \begin{pmatrix} 1 \\ 2 \end{pmatrix} + \begin{pmatrix} 3 \\ 2 \end{pmatrix} \right\} \right)$$
$$= \text{conv}\left( \left\{ \begin{pmatrix} 3 \\ 2 \end{pmatrix}, \begin{pmatrix} 4 \\ 3 \end{pmatrix}, \begin{pmatrix} 3 \\ 3 \end{pmatrix}, \begin{pmatrix} 4 \\ 4 \end{pmatrix} \right\} \right)$$

which is sketched in Figure 3.5. Observe that none of the points are redundant, hence they are all extreme.

(iii) A minor generalization of the procedure from Exercise 3.7 gives
$$\begin{aligned} &\text{conv}(\{x_1, \ldots, x_m\}) + \text{conv}(\{y_1, y_2\}) \\ &\quad = \text{conv}(\{x_1 + y_1, \ldots, x_m + y_1, x_1 + y_2, \ldots, x_m + y_2\}) \end{aligned} \tag{3.6}$$
for $x_1, \ldots, x_m, y_1, y_2 \in \mathbb{R}^n$. From this we infer
$$Q = \text{conv}\left( \left\{ \begin{pmatrix} 6 \\ 3 \end{pmatrix}, \begin{pmatrix} 6 \\ 4 \end{pmatrix}, \begin{pmatrix} 7 \\ 3 \end{pmatrix}, \begin{pmatrix} 7 \\ 5 \end{pmatrix}, \begin{pmatrix} 8 \\ 4 \end{pmatrix}, \begin{pmatrix} 8 \\ 5 \end{pmatrix} \right\} \right),$$
where we have excluded the redundant point $(7, 4)$ by considering Figure 3.5. Furthermore, from the figure we see that
$$Q = \{(x, y) \in \mathbb{R}^2 \mid 6 \le x \le 8, \ 3 \le y \le 5, \ x - 4 \le y \le x - 2\}.$$

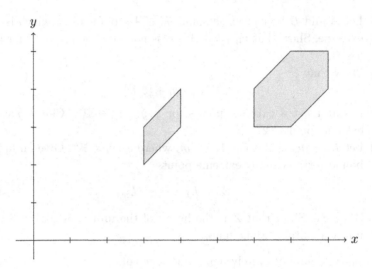

**Figure 3.5:** The sets $P$ and $Q$.

(iv) Since $u_0$ is not extreme in $A$, we can find $x, y \in A \setminus \{u_0\}$ with $u_0 \in [x, y]$. Now let $b \in B$ and note that $u_0 + b \in [x + b, y + b]$ by (i). As $x + b, y + b \in A + B$ and none of them are equal to $u_0 + b$, we have the result.

(v) Exercise 3.7 tells us that $T = \mathrm{conv}(\{x + z, x + w, y + z, y + w\})$, and then it follows from Exercise 3.13(i) that an extreme point of $T$ must be one of those four points.

To prove that $T$ cannot have three extreme points, we start by noting that the extreme points are exactly the non-redundant generators (cf. Exercise 3.13(ii)). Thus, if $T$ has three or less extreme points we may without loss of generality assume that $x + z$ is redundant and write

$$x + z = \lambda_1(x + w) + \lambda_2(y + z) + \lambda_3(y + w) \qquad (3.7)$$

for some $\lambda_1, \lambda_2, \lambda_3 \geq 0$ satisfying $\lambda_1 + \lambda_2 + \lambda_3 = 1$. If $z = w$, $T$ has two extreme points by (i), so we assume $\lambda_1 \neq 1$. By rearranging (3.7) we find $x - y = \alpha(w - z)$, where $\alpha = (\lambda_1 + \lambda_3)(\lambda_2 + \lambda_3)^{-1}$. The equations

$$(y + z) - (x + w) = (1 + \alpha)(z - w)$$
$$(y + w) - (x + w) = \alpha(z - w)$$

imply that $x + w, y + z$, and $y + w$ are affinely dependent by Proposition 2.9(2). Consequently, $\text{conv}(\{x + w, y + z, y + w\})$ is a line segment and only has two extreme points.

It is possible for $T$ to have two extreme points. For instance, it is the case if $z = w = 0$ and $x \neq y$.

(vi) Inductively, (3.6) shows that $Z$ is generated by at most $2^m$ elements, which will be an upper bound for the number of extreme points cf. Exercise 3.13(i). We will now show that $z \in Z$ if and only if $z = A\lambda + b$, where $A$ is an $n \times m$ matrix, $b \in \mathbb{R}^n$, and $\lambda \in [0,1]^m$. Let $i \in \{1, \ldots, m\}$ and write

$$L_i = \{(1 - \lambda_i)u_i + \lambda_i v_i \mid \lambda_i \in [0,1]\} = \{\lambda_i a_i + b_i \mid \lambda_i \in [0,1]\}$$

where $a_i = v_i - u_i \in \mathbb{R}^n$ and $b_i = u_i \in \mathbb{R}^n$. We conclude that

$$
\begin{aligned}
Z &= \{l_1 + \cdots + l_m \mid l_1 \in L_1, \ldots, l_m \in L_m\} \\
&= \{(\lambda_1 a_1 + b_1) + \cdots + (\lambda_m a_m + b_m) \mid \lambda_1, \ldots, \lambda_m \in [0,1]\} \\
&= \{A\lambda + b \mid \lambda \in [0,1]^m\}
\end{aligned}
$$

where $a_i$ is the $i$-th column in $A$ and $b = b_1 + \cdots + b_m$.

$\star \quad \star \quad \star$

**Exercise 3.16.** Give an example of a non-convex cone.

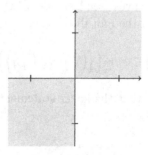

**Figure 3.6:** The non-convex cone $K$ from Solution 3.16.

**Solution 3.16.** The non-convex cone $K$ is depicted in Figure 3.6 where

$$K = \{(x,y) \in \mathbb{R}^2 \mid x \geq 0, y \geq 0\} \cup \{(x,y) \in \mathbb{R}^2 \mid x \leq 0, y \leq 0\}.$$

$\star \quad \star \quad \star$

**Exercise 3.17.** Prove in detail that

$$C = \{(x, y, z) \in \mathbb{R}^3 \mid z \geq 0, \, x^2 + y^2 \leq z^2\}$$

is a convex cone. Is $C$ finitely generated?

**Solution 3.17.** The key observation in this exercise is to note that

$$C = \left\{(x, y, z) \in \mathbb{R}^3 \,\middle|\, \left\|\begin{pmatrix} x \\ y \end{pmatrix}\right\| \leq z\right\}.$$

Next, applying the triangle inequality (Theorem A.2), it follows that $C$ is a convex subset. An illustration of $C$ is given in [U.C., Figure 3.11] and this indicates that it is not finitely generated. If we assume that $C$ is finitely generated we may find $n \in \mathbb{N}$ and $v_1, \ldots, v_n \in \mathbb{R}^2$ such that

$$C = \text{cone}\left(\left\{\begin{pmatrix} v_1 \\ 1 \end{pmatrix}, \ldots, \begin{pmatrix} v_n \\ 1 \end{pmatrix}\right\}\right). \tag{3.8}$$

If we can show that the assumption implies

$$\text{conv}(\{v_1, \ldots, v_n\}) = \left\{\begin{pmatrix} x \\ y \end{pmatrix} \in \mathbb{R}^2 \,\middle|\, x^2 + y^2 \leq 1\right\},$$

Exercise 3.13(i) gives that the unit disc only has finitely many extreme points which contradicts Exercise 3.14. One of the inclusions follows from the fact that $v_1, \ldots, v_n$ all lie on the unit circle, since $(v_1, 1), \ldots, (v_n, 1) \in C$, and that the unit disc is convex (see Exercise 3.14). For the other inclusion we note that if $(x, y)$ is on the unit disc, $(x, y, 1) \in C$ and consequently,

$$\begin{pmatrix} x \\ y \\ 1 \end{pmatrix} \in \text{cone}\left(\left\{\begin{pmatrix} v_1 \\ 1 \end{pmatrix}, \ldots, \begin{pmatrix} v_n \\ 1 \end{pmatrix}\right\}\right)$$

by (3.8). Finally, we note that the latter statement is equivalent of saying $(x, y) \in \text{conv}(\{v_1, \ldots, v_n\})$.

⋆    ⋆    ⋆

**Exercise 3.18.** Prove in detail that the recession cone $\text{rec}(C)$ defined in [U.C., (3.3)] is a convex cone, where $C$ is a convex subset. Perhaps the identity

$$x + (n + \lambda)d = (1 - \lambda)(x + nd) + \lambda(x + (n + 1)d)$$

might come in handy.

**Solution 3.18.** Let $d_1, d_2 \in \text{rec}(C)$ and $\lambda \in [0, 1]$. We need to show that $x + (1 - \lambda)d_1 + \lambda d_2 \in C$ for an arbitrary $x \in C$. For this we notice that

$$x + (1 - \lambda)d_1 + \lambda d_2 = (1 - \lambda)(x + d_1) + \lambda(x + d_2) \in C,$$

since $x + d_1, x + d_2 \in C$ and $C$ is a convex set. First, in order to show the cone property, we note that $nd \in \text{rec}(C)$ if $d \in \text{rec}(C)$ (argue by induction using the relation $x + nd = (x + (n-1)d) + d)$. Now let $\mu \geq 0$ and $d \in \text{rec}(C)$. We can find $n \in \{0, 1, 2, \dots\}$ and $\lambda \in [0, 1]$ such that $\mu = n + \lambda$, and if we combine this with the hint given in the exercise we get

$$x + \mu d = (1 - \lambda)(x + nd) + \lambda(x + (n + 1)d).$$

We have that $x + nd, x + (n+1)d \in C$ and since $C$ is convex, it follows that $x + \mu d \in C$. Consequently, $\text{rec}(C)$ is a cone.

$\star \quad \star \quad \star$

**Exercise 3.19.** What is the recession cone of a bounded convex subset?

**Solution 3.19.** Let $C$ be a convex set. By definition, $0 \in \text{rec}(C)$. Now suppose that we may choose $d \in \text{rec}(C)$ with $d_i \neq 0$ for some $i \in \{1, \dots, n\}$. Since $\text{rec}(C)$ is a cone we know that for any $x \in C$, then $x + \lambda d \in C$ for $\lambda \geq 0$. As we have that

$$|x + \lambda d| = \sqrt{(x_1 + \lambda d_1)^2 + \cdots + (x_n + \lambda d_n)^2} \geq \sqrt{(x_i + \lambda d_i)^2} = |x_i + \lambda d_i|$$

and $|x_i + \lambda d_i| \to \infty$ as $\lambda \to \infty$, $C$ must be unbounded. Therefore the recession cone of a bounded convex set is equal to $\{0\}$.

$\star \quad \star \quad \star$

**Exercise 3.20.** Can you give an example of an unbounded convex subset $C \subseteq \mathbb{R}^2$ with $\text{rec}(C) = \{0\}$?

**Solution 3.20.** An example is given by

$$C = \{(x, y) \in \mathbb{R}^2 | \ 0 \leq x < 1\} \cup \{(1, 0)\}.$$

One can check this by considering $d \neq 0$ and then verify that we can find $(x, y) \in C$ such that $(x, y) + d \notin C$. A geometric procedure for verifying this can be found in Figure 3.7 where the different arrows (corresponding to different $d$) point to values of $(x, y) + d$ for certain critical $(x, y)$.

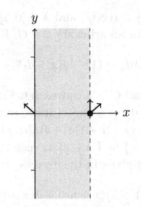

**Figure 3.7:** The set $C = \{(x,y) \in \mathbb{R}^2 \mid 0 \le x < 1\} \cup \{(1,0)\}$ is an example of an unbounded convex set with $\mathrm{rec}(C) = \{0\}$.

$$\star \quad \star \quad \star$$

**Exercise 3.21.** Let

$$C = \mathrm{cone}\left(\left\{\begin{pmatrix}2\\1\end{pmatrix}, \begin{pmatrix}1\\2\end{pmatrix}\right\}\right).$$

(i) Show that

$$C^\circ = \mathrm{cone}\left(\left\{\begin{pmatrix}1\\-2\end{pmatrix}, \begin{pmatrix}-2\\1\end{pmatrix}\right\}\right).$$

(ii) Suppose that

$$C = \mathrm{cone}\left(\left\{\begin{pmatrix}a\\c\end{pmatrix}, \begin{pmatrix}b\\d\end{pmatrix}\right\}\right),$$

where

$$\begin{pmatrix} a & b \\ c & d \end{pmatrix}$$

is an invertible matrix. How do you compute $C^\circ$?

**Solution 3.21.**     (i) It follows from the proof of Proposition 3.12 that

$$C^\circ = \left\{\alpha \in \mathbb{R}^2 \;\middle|\; \alpha^t\begin{pmatrix}2\\1\end{pmatrix} \le 0, \; \alpha^t\begin{pmatrix}1\\2\end{pmatrix} \le 0\right\}.$$

Now let $\alpha \in \mathbb{R}^2$. Since $(1, -2)$ and $(-2, 1)$ are linearly independent we may find unique $\lambda_1, \lambda_2 \in \mathbb{R}$, such that

$$\lambda_1\begin{pmatrix}1\\-2\end{pmatrix} + \lambda_2\begin{pmatrix}-2\\1\end{pmatrix} = \alpha.$$

Using this relation we compute

$$\alpha^t\begin{pmatrix} 2 \\ 1 \end{pmatrix} = -3\lambda_2 \quad \text{and} \quad \alpha^t\begin{pmatrix} 1 \\ 2 \end{pmatrix} = -3\lambda_1,$$

which are non-positive if and only if $\lambda_1, \lambda_2 \geq 0$. Thus,

$$C^o = \text{cone}\left(\left\{\begin{pmatrix} 1 \\ -2 \end{pmatrix}, \begin{pmatrix} -2 \\ 1 \end{pmatrix}\right\}\right).$$

(ii) As the associated matrix is invertible, it has a non-zero determinant. Assume (as was the case in (i)) that

$$\det \begin{pmatrix} a & b \\ c & d \end{pmatrix} > 0.$$

In the same way as in (i) we verify that

$$C^o = \text{cone}\left(\left\{\begin{pmatrix} c \\ -a \end{pmatrix}, \begin{pmatrix} -d \\ b \end{pmatrix}\right\}\right).$$

If the determinant is negative note that

$$C = \text{cone}\left(\left\{\begin{pmatrix} b \\ d \end{pmatrix}, \begin{pmatrix} a \\ c \end{pmatrix}\right\}\right),$$

and the associated matrix in this setup has a positive determinant. Consequently, we can apply the result above and obtain that

$$C^o = \text{cone}\left(\left\{\begin{pmatrix} d \\ -b \end{pmatrix}, \begin{pmatrix} -c \\ a \end{pmatrix}\right\}\right).$$

$$\star \quad \star \quad \star$$

**Exercise 3.22.** The vector

$$v = \begin{pmatrix} 7/4 \\ 19/8 \end{pmatrix}$$

is the convex combination

$$\frac{1}{8}\begin{pmatrix} 1 \\ 1 \end{pmatrix} + \frac{1}{8}\begin{pmatrix} 1 \\ 2 \end{pmatrix} + \frac{1}{4}\begin{pmatrix} 2 \\ 2 \end{pmatrix} + \frac{1}{2}\begin{pmatrix} 2 \\ 3 \end{pmatrix}$$

of four vectors in $\mathbb{R}^2$. Use the method outlined in Example 3.16 to answer the following questions.

(i) Is $v$ in the convex hull of three of the four vectors?
(ii) Is $v$ in the convex hull of two of the four vectors?

**Solution 3.22.**     (i) In line with the procedure from Example 3.16 we start by noticing the linear dependence

$$\begin{pmatrix}2\\2\\1\end{pmatrix}+\begin{pmatrix}1\\2\\1\end{pmatrix}-\begin{pmatrix}1\\1\\1\end{pmatrix}-\begin{pmatrix}2\\3\\1\end{pmatrix}=\begin{pmatrix}0\\0\\0\end{pmatrix},$$

and from this we deduce the equality

$$\begin{pmatrix}7/4\\19/8\\1\end{pmatrix}=(\tfrac{1}{8}+\theta)\begin{pmatrix}1\\1\\1\end{pmatrix}+(\tfrac{1}{8}-\theta)\begin{pmatrix}2\\1\\1\end{pmatrix}+(\tfrac{1}{4}-\theta)\begin{pmatrix}2\\2\\1\end{pmatrix}+(\tfrac{1}{2}+\theta)\begin{pmatrix}2\\3\\1\end{pmatrix}. \qquad (3.9)$$

Since the coefficients sum to one, choosing the largest $\theta$ such that all the scalars are non-negative gives $v$ as a convex combination of three vectors. This is achieved for $\theta = 1/8$ and by inserting in (3.9), we get

$$v = \tfrac{1}{4}\begin{pmatrix}1\\1\end{pmatrix}+\tfrac{1}{8}\begin{pmatrix}2\\2\end{pmatrix}+\tfrac{5}{8}\begin{pmatrix}2\\3\end{pmatrix}. \qquad (3.10)$$

The three vectors in (3.10) are affinely independent and we cannot continue the algorithm.

(ii) The question is whether $v$ is on a line between any two of the four generating vectors. Since $19/8 > 2$, $(2,3)$ has to be one of them, and $7/4 < 2$ so $(2,2)$ cannot be the second point. To see that $v \notin \mathrm{conv}(\{(1,2),(2,3)\})$ note that

$$\lambda_1\begin{pmatrix}1\\2\\1\end{pmatrix}+\lambda_2\begin{pmatrix}2\\3\\1\end{pmatrix}=\begin{pmatrix}7/4\\19/8\\1\end{pmatrix}$$

has no (non-negative) solution. The same type of calculations show that $v \notin \mathrm{conv}(\{(1,1),(2,3)\})$.

These conclusions are in line with Figure 3.8.

$$\star \quad \star \quad \star$$

**Exercise 3.23.** Let $C_1, C_2, C_3, C_4$ be convex subsets of $\mathbb{R}^2$, such that any three of them have non-empty intersection i.e.,

$$C_1 \cap C_2 \cap C_3 \neq \emptyset$$
$$C_1 \cap C_2 \cap C_4 \neq \emptyset$$
$$C_1 \cap C_3 \cap C_4 \neq \emptyset$$
$$C_2 \cap C_3 \cap C_4 \neq \emptyset.$$

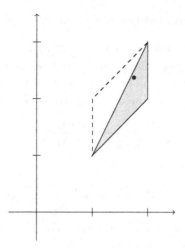

**Figure 3.8:** The convex hull before and after the iteration and the point
$v = (7/4, 19/8)$.

(i) Show that for $v_1, v_2, v_3, v_4 \in \mathbb{R}^2$ there are $\lambda_1, \lambda_2, \lambda_3, \lambda_4 \in \mathbb{R}$, not all
zero, such that

$$\lambda_1 v_1 + \lambda_2 v_2 + \lambda_3 v_3 + \lambda_4 v_4 = 0$$
$$\lambda_1 + \lambda_2 + \lambda_3 + \lambda_4 = 0 \,.$$

(ii) By assumption there exists

$$v_i \in \bigcap_{\substack{j=1 \\ j \neq i}}^{4} C_j$$

for $i = 1, \ldots, 4$. Suppose that with the notation in (i) we have

$$\lambda_1, \lambda_2 \geq 0$$
$$\lambda_3, \lambda_4 \leq 0 \,.$$

Prove that

$$\frac{\lambda_1}{\lambda_1 + \lambda_2} v_1 + \frac{\lambda_2}{\lambda_1 + \lambda_2} v_2 \in C_1 \cap C_2 \cap C_3 \cap C_4 \,.$$

(iii) Prove now in general that $C_1 \cap C_2 \cap C_3 \cap C_4 \neq \emptyset$.

(iv) Prove that if $C_1, \ldots, C_m$ are convex subsets of $\mathbb{R}^2$ where any three
of them have non-empty intersection, then $C_1 \cap \cdots \cap C_m \neq \emptyset$.

(v) What is a natural generalization from $\mathbb{R}^2$ to $\mathbb{R}^n$ of the result in (iv)?

**Solution 3.23.**    (i) By Exercise 2.10, $v_1, v_2, v_3$, and $v_4$ are affinely dependent and hence, the statement follows by Proposition 2.9(3).

(ii) According to (i) we can be sure that $\lambda_1 + \lambda_2 > 0$. It is assumed that $v_1, v_2 \in C_3 \cap C_4$ and the convexity of $C_3 \cap C_4$ ensures

$$\frac{\lambda_1}{\lambda_1 + \lambda_2} v_1 + \frac{\lambda_2}{\lambda_1 + \lambda_2} v_2 \in C_3 \cap C_4.$$

Using (i) gives

$$\frac{\lambda_1}{\lambda_1 + \lambda_2} v_1 + \frac{\lambda_2}{\lambda_1 + \lambda_2} v_2 = \frac{\lambda_3}{\lambda_3 + \lambda_4} v_3 + \frac{\lambda_4}{\lambda_3 + \lambda_4} v_4 \in C_1 \cap C_2$$

by assumption and the convexity of $C_1 \cap C_2$, and we conclude that

$$\frac{\lambda_1}{\lambda_1 + \lambda_2} v_1 + \frac{\lambda_2}{\lambda_1 + \lambda_2} v_2 \in C_1 \cap C_2 \cap C_3 \cap C_4.$$

(iii) Combining (i) and (ii) we assume without loss of generality that we are in one of the two cases $\lambda_1 \leq 0$ and $\lambda_2, \lambda_3, \lambda_4 \geq 0$ or $\lambda_1 \geq 0$ and $\lambda_2, \lambda_3, \lambda_4 \leq 0$. In both cases we have that

$$v_1 = \frac{\lambda_2}{\lambda_2 + \lambda_3 + \lambda_4} v_2 + \frac{\lambda_3}{\lambda_2 + \lambda_3 + \lambda_4} v_3 + \frac{\lambda_4}{\lambda_2 + \lambda_3 + \lambda_4} v_4 \in C_1$$

by the convexity of $C_1$. It is assumed that $v_1 \in C_2 \cap C_3 \cap C_4$, so we conclude that $v_1 \in C_1 \cap C_2 \cap C_3 \cap C_4$.

(iv) The case where $m = 4$ is verified by (i), (ii) and (iii). For general $m \geq 5$, assume that we can choose

$$v_i \in \cap_{j=1, j \neq i}^{m} C_j$$

for $i = 1, \ldots, m$. These $m$ points are affinely dependent and hence, we can choose $\lambda_1, \ldots, \lambda_m \in \mathbb{R}$, not all zero, such that $\lambda_1 v_1 + \cdots + \lambda_m v_m = 0$ and $\lambda_1 + \cdots + \lambda_m = 0$. Assume without loss of generality that $\lambda_1, \ldots, \lambda_k > 0$ and $\lambda_{k+1}, \ldots, \lambda_m \leq 0$ for some $1 \leq k < m$. Then we have the relation

$$\frac{\lambda_1}{\lambda_1 + \cdots + \lambda_k} v_1 + \cdots + \frac{\lambda_k}{\lambda_1 + \cdots + \lambda_k} v_k$$

$$= \frac{\lambda_{k+1}}{\lambda_{k+1} + \cdots + \lambda_m} v_{k+1} + \cdots + \frac{\lambda_m}{\lambda_{k+1} + \cdots + \lambda_m} v_m.$$

This shows that $C_1 \cap \cdots \cap C_m \neq \emptyset$, and now the result follows by induction.

(v) Let $C_1, \ldots, C_m \subseteq \mathbb{R}^n$, $m \geq n+2$, be convex sets such that any $n+1$ of the sets have a non-empty intersection. Then you can prove that $C_1 \cap \cdots \cap C_m \neq \emptyset$ (see Exercise 3.25).

$$\star \quad \star \quad \star$$

**Exercise 3.24.** Let $S$ be a subset of $\mathbb{R}^n$ containing at least $n+2$ points. Prove that there exists subsets $S_1, S_2 \subseteq S$, such that

(i) $S_1 \cap S_2 = \emptyset$
(ii) $S_1 \cup S_2 = S$
(iii) $\operatorname{conv}(S_1) \cap \operatorname{conv}(S_2) \neq \emptyset$.

Hint: write down an affine dependence between $v_1, \ldots, v_{n+2} \in S$. This result is called Radon's theorem.

**Solution 3.24.** Let $v_1, \ldots, v_{n+2} \in S \subseteq \mathbb{R}^n$. Combining Exercise 2.10 and Proposition 2.9(3) there exists $\lambda_1, \ldots, \lambda_{n+2} \in \mathbb{R}$, not all zero, such that

$$\lambda_1 v_1 + \cdots + \lambda_{n+2} v_{n+2} = 0 \qquad (3.11)$$
$$\lambda_1 + \cdots + \lambda_{n+2} = 0. \qquad (3.12)$$

By possibly changing the order in which we enumerate $v_1, \ldots, v_{n+2}$ we can assume that $\lambda_1, \ldots, \lambda_k > 0$ and $\lambda_{k+1}, \ldots, \lambda_{n+2} \leq 0$ for some $1 \leq k \leq n+1$. Then we define

$$S_1 = \{v_1, \ldots, v_k\}$$
$$S_2 = S \setminus \{v_1, \ldots, v_k\}.$$

Condition (i) and (ii) follow straight from these definitions. Using (3.12) to write (3.11) as

$$\frac{\lambda_1}{\lambda_1 + \cdots + \lambda_k} v_1 + \cdots + \frac{\lambda_k}{\lambda_1 + \cdots + \lambda_k} v_k$$
$$= \frac{\lambda_{k+1}}{\lambda_{k+1} + \cdots + \lambda_m} v_{k+1} + \cdots + \frac{\lambda_{n+2}}{\lambda_{k+1} + \cdots + \lambda_{n+2}} v_{n+2}$$

shows that $\operatorname{conv}(S_1) \cap \operatorname{conv}(S_2) \neq \emptyset$.

$$\star \quad \star \quad \star$$

**Exercise 3.25.** Use the result in Exercise 3.24 to give a complete proof of the natural generalization alluded to in Exercise (v) (called Helly's theorem).

**Solution 3.25.** We will prove that $n + 2$ or more convex subsets in $\mathbb{R}^n$, where any intersection of $n + 1$ of the subsets is non-empty, share at least one element. In order to prove this we start by showing that the result holds if we require that every $m - 1 \geq n + 1$ convex sets have a non-empty intersection.

Let $C_1, \ldots, C_m \subseteq \mathbb{R}^n$ be convex sets where $m \geq n + 2$ and such that any intersection of $m - 1$ of them are non-empty. Then the intersection of all $m$ convex sets is non-empty. To prove this statement we start by letting

$$v_i \in \bigcap_{\substack{j=1 \\ j \neq i}}^{m} C_j$$

for $i = 1, \ldots, m$. Then Exercise 3.24 gives, by possibly changing the order, that there exists $k$ such that

$$\operatorname{conv}(\{v_1, \ldots, v_k\}) \cap \operatorname{conv}(\{v_{k+1}, \ldots, v_m\}) \neq \emptyset. \tag{3.13}$$

Note that

$$\operatorname{conv}(\{v_1, \ldots, v_k\}) \subseteq \bigcap_{j=k+1}^{m} C_j \quad \text{and} \quad \operatorname{conv}(\{v_{k+1}, \ldots, v_m\}) \subseteq \bigcap_{j=1}^{k} C_j$$

by Proposition 3.4. As a consequence,

$$\operatorname{conv}(\{v_1, \ldots, v_k\}) \cap \operatorname{conv}(\{v_{k+1}, \ldots, v_m\}) \subseteq \bigcap_{j=1}^{m} C_j,$$

and the result follows from (3.13).

The more general result now follows by induction.

$$\star \quad \star \quad \star$$

**Exercise 3.26.** Let $e_1, e_2, e_3$ denote the canonical basis vectors of $\mathbb{R}^3$ and let

$$C = \operatorname{conv}(\{e_1, -e_1, e_2, -e_2, e_3, -e_3\}).$$

Verify that

$$\left(\tfrac{1}{3}, \tfrac{1}{5}, \tfrac{1}{7}\right) \in C$$

by writing down a convex linear combination.

**Solution 3.26.** We are looking for a non-negative solution to

$$
\begin{pmatrix}
1 & -1 & 0 & 0 & 0 & 0 \\
0 & 0 & 1 & -1 & 0 & 0 \\
0 & 0 & 0 & 0 & 1 & -1 \\
1 & 1 & 1 & 1 & 1 & 1
\end{pmatrix}
\lambda =
\begin{pmatrix}
1/3 \\
1/5 \\
1/7 \\
1
\end{pmatrix}.
$$

A solution is

$$
\lambda = \left( \tfrac{1}{3}, 0, \tfrac{1}{5}, 0, \tfrac{32}{105}, \tfrac{17}{105} \right).
$$

This can be found by letting two entrances in $\lambda$ be zero, solving the system by inverting a square matrix and then checking for non-negativity.

# Chapter 4

# Polyhedra

## 4.1 Introduction

The exercises focus mostly on helping the understanding of the more theoretical chapter on polyhedra. First it is edges and related terminology that is in focus. Later we will use Farkas's lemma and Weyl's theorem. Here it is interesting to note that Farkas's lemma can be seen as an equivalent condition for a point to lie in a finitely generated cone. The last exercises, which are concerned with Markov chains, doubly stochastic matrices and perfect pairings, are more of independent interest.

Let $P = \{x \in \mathbb{R}^d \mid Ax \le b\}$, where $A$ is an $m \times d$ matrix and $b \in \mathbb{R}^m$. For a subset $I \subseteq \{1, \ldots, m\}$, we define

$$P_I := \{x \in P \mid A_I x = b_I\} \tag{4.1}$$

where $A_I$ is the submatrix with rows in $I$ and similarly $b_I$ is the subvector of $b$ with coordinates in $I$.

**Lemma 4.2.** *Suppose that $I \subseteq \{1, \ldots, m\}$. If $P_I \ne \emptyset$, then $P_I$ is an exposed face of $P$.*

Let

$$I(z) = \{i \in \{1, \ldots, m\} \mid a_i z = b_i\} \tag{4.2}$$

for $z \in \mathbb{R}^d$ and put $A_z = A_{I(z)}$.

**Proposition 4.3.** *If $F$ is a non-empty face of $P$, then $F = P_I$ for some $I \subseteq \{1, \ldots, m\}$. If $P_I$ is a non-empty face with $I(z) = I$ for some $z \in P_I$, then*

$$\dim P_I = d - \operatorname{rk} A_I.$$

**Definition 4.11.** Suppose that $A$ is an $m \times d$ matrix and $R$ a $d \times n$ matrix. Then $(A, R)$ is called a *double description pair* if

$$\{x \in \mathbb{R}^d \mid Ax \leq 0\} = \{R\lambda \mid \lambda \in \mathbb{R}^n, \, \lambda \geq 0\}.$$

**Theorem 4.12 (Weyl).** *For every matrix $R$, there exists a matrix $A$, such that $(A, R)$ is a double description pair.*

**Lemma 4.14 (Farkas).** *Let $A$ be an $m \times n$ matrix and $b \in \mathbb{R}^m$. Then precisely one of the following two conditions hold.*

(1) *The system $Ax = b$ of linear equations is solvable with $x \geq 0$.*
(2) *There exists $y \in \mathbb{R}^m$ such that*

$$y^t A \leq 0 \quad \text{and} \quad y^t b > 0.$$

Recall that a $P \subseteq \mathbb{R}^n$ is a polyhedron if it can be written as the solution to a set of inequalities (see Definition 1.3). Furthermore, a polytope is a convex hull that can be generated by finitely many vectors (see Definition 3.3).

**Theorem 4.23 (Minkowski and Weyl).** *A subset $P \subseteq \mathbb{R}^n$ is a polyhedron if and only if*

$$P = C + Q,$$

*where $C$ is a finitely generated cone and $Q$ a polytope.*

**Theorem 4.24.** *Let $P = \{x \in \mathbb{R}^n \mid Ax \leq b\}$ be a polyhedron with $\text{ext}(P) \neq \emptyset$. Then*

$$P = \text{conv}(\text{ext}(P)) + \text{rec}(P).$$

A doubly stochastic matrix is a square matrix with non-negative entries, and row and column sum equal to one. A permutation matrix is a doubly stochastic matrix with entries equal to 0 or 1.

**Theorem 4.26 (Birkhoff, von Neumann).** *A doubly stochastic matrix is a convex linear combination of permutation matrices.*

## 4.2    Exercises and solutions

**Exercise 4.1.** Let $P = \{x \in \mathbb{R}^3 \mid Ax \leq b\}$ be a polyhedron in $\mathbb{R}^3$ and suppose that $z \notin P$. Give a criterion detecting if an extreme point $x \in \text{ext}(P)$ is visible from $z$.

**Solution 4.1.** A natural way to interpret visibility is that $x$ is not "hidden" behind other points in $P$ when viewed from $z$. In mathematical terms this requirement could be formulated as

$$(1 - \lambda)x + \lambda z \notin P \quad \text{for every } \lambda \in (0, 1). \tag{4.3}$$

In our concrete setup we claim that this visibility condition is equivalent to the existence of an $i \in I(x)$ (cf. (4.2)) such that $(Az)_i > b_i$. Let us do the proof.

Suppose that (4.3) is satisfied. As $z \notin P$ we must have $(Az)_i > b_i$ for some $i \in \{1, \ldots, m\}$. If we had $(Ax)_i < b_i$ for every such $i$, $(1-\lambda)x + \lambda z \in P$ for $\lambda \in (0, 1)$ sufficiently small, contradicting the assumption. Consequently, we can choose $i \in I(x)$ with $(Az)_i > b_i$. On the other hand, if there exists $i \in I(x)$ such that $(Az)_i > b_i$, let $\lambda \in (0, 1)$ and consider $v = (1 - \lambda)x + \lambda z$ so that $Av = (1 - \lambda)b + \lambda Az$. This means $(Av)_i > b_i$ implying $v \notin P$.

$\star \quad \star \quad \star$

**Exercise 4.2.** Compute the edges of $P$ in Example 4.6 using Proposition 4.3.

**Solution 4.2.** Due to Lemma 4.2 and Proposition 4.3 we consider submatrices of

$$A = \begin{pmatrix} -1 & -1 \\ 2 & -1 \\ -1 & 2 \\ 1 & 2 \end{pmatrix}$$

with rank one. These are exactly the different rows of $A$ and thus, we find the edges

$$P_1 = \{v \in P \mid (1, 1)v = 0\} = \{(v_1, v_2) \in \mathbb{R}^2 \mid v_1 + v_2 = 0, -\tfrac{1}{3} \leq v_1 \leq \tfrac{1}{3}\}$$
$$P_2 = \{v \in P \mid (2, -1)v = 1\} = \{(v_1, v_2) \in \mathbb{R}^2 \mid 2v_1 - v_2 = 1, \tfrac{1}{3} \leq v_1 \leq \tfrac{4}{5}\}$$
$$P_3 = \{v \in P \mid (-1, 2)v = 1\} = \{(v_1, v_2) \in P \mid -v_1 + 2v_2 = 1, -\tfrac{1}{3} \leq v_1 \leq \tfrac{1}{2}\}$$
$$P_4 = \{v \in P \mid (1, 2)v = 2\} = \{(v_1, v_2) \in P \mid v_1 + 2v_2 = 2, \tfrac{1}{2} \leq v_1 \leq \tfrac{4}{5}\}.$$

To conclude that these sets are one-dimensional, one verifies the existence of the $z$ given in Proposition 4.3.

$\star \quad \star \quad \star$

**Exercise 4.3.** Let $Q = \text{conv}(\{v_1, \dots, v_m\})$, where $v_i \in \mathbb{R}^d$. Suppose that $Q = \{x \in \mathbb{R}^d \mid Ax \le b\}$. Prove that $E = \text{conv}(\{v_i, v_j\})$ is an edge of $Q$ if $\text{rk} A_z = d - 1$, where $z = \frac{1}{2}(v_i + v_j)$.

**Solution 4.3.** First, let $I = I(z)$ and note that $z \in Q_I$. By Lemma 4.2, $Q_I$ is an exposed face, and then it follows by Proposition 4.3 that

$$\dim Q_I = d - \text{rk} \, A_I = 1,$$

which means $Q_I$ is an edge of $Q$. Thus, it suffices to show that $\text{conv}(\{v_i, v_j\}) = Q_I$. It must be the case that $A_I v_i = A_I v_j = b_I$ (since $v_i, v_j \in Q$ and $z = (v_i + v_j)/2$) and consequently, $A_I x = b_I$ for every $x \in \text{conv}(\{v_i, v_j\})$. Therefore $\text{conv}(\{v_i, v_j\}) \subseteq Q_I$. To obtain equality, we need to assume that $v_i, v_j \in \text{ext}(Q)$. Since $v_i \ne v_j$ it follows that $Q_I \subseteq \text{aff}(\{v_i, v_j\})$. Hence, if there exists a point $y \in Q_I \setminus \text{conv}(\{v_i, v_j\})$ it must be of the form $y = (1 - t)v_i + tv_j$ for a given $t \in \mathbb{R} \setminus [0, 1]$. We may assume that $t > 1$. This gives the relation

$$v_j = \tfrac{t-1}{t}v_i + \tfrac{1}{t}y,$$

from which it follows that $v_j \in \{v_i, y\}$ since $v_j$ is an extreme point of $Q$ and thus, we have a contradiction. This implies that $\text{conv}(\{v_i, v_j\}) = Q_I$.

⋆ ⋆ ⋆

**Exercise 4.4.** Show using the lemma of Farkas that

(i) The equation

$$\begin{pmatrix} 1 & 2 & 3 \\ 3 & 1 & 5 \end{pmatrix} \begin{pmatrix} x \\ y \\ z \end{pmatrix} = \begin{pmatrix} 5 \\ 2 \end{pmatrix}$$

is unsolvable with $x \ge 0, y \ge 0$ and $z \ge 0$.

(ii) The equation

$$\begin{pmatrix} 1 & 2 & 3 & 1 \\ 3 & 1 & 5 & 1 \\ 1 & 2 & 1 & 1 \end{pmatrix} \begin{pmatrix} x \\ y \\ z \\ w \end{pmatrix} = \begin{pmatrix} 2 \\ 1 \\ 1 \end{pmatrix}$$

is unsolvable with $x \ge 0, y \ge 0, z \ge 0$ and $w \ge 0$.

**Solution 4.4.** (i) From Lemma 4.14 (Farkas) we know the system is unsolvable for $x \geq 0$, $y \geq 0$, and $z \geq 0$ if there exists $v \in \mathbb{R}^2$ such that

$$v^t \begin{pmatrix} 1 & 3 \\ 2 & 1 \\ 3 & 5 \end{pmatrix} \leq 0 \text{ and } v^t \begin{pmatrix} 5 \\ 2 \end{pmatrix} > 0. \tag{4.4}$$

Using Figure 4.1 we see that $v = (-1, 2)$ satisfies (4.4).

(ii) We are now looking for $v \in \mathbb{R}^3$ such that

$$v^t \begin{pmatrix} 1 & 2 & 3 & 1 \\ 3 & 1 & 5 & 1 \\ 1 & 2 & 1 & 1 \end{pmatrix} \leq 0 \text{ and } v^t \begin{pmatrix} 2 \\ 1 \\ 1 \end{pmatrix} > 0. \tag{4.5}$$

This problem is not easily drawn and we will instead do some heuristic reasoning. We start by noting that one of the entries in $v$ must be positive since $2v_1 + v_2 + v_3 > 0$, and that the 2 in front of $v_1$ could make this coordinate a good candidate. On the other hand, every entry in the $3 \times 4$ matrix above is positive, so some of the entries in $v$ must be negative. From this we guess (and verifies) that $v = (2, -2, -1)$ satisfies (4.5), and Lemma 4.14 (Farkas) gives the desired result.

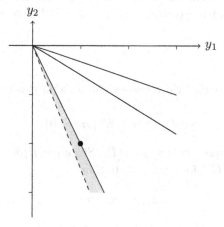

**Figure 4.1:** The elements such that (4.4) is satisfied together with the point $(1, -2)$.

★  ★  ★

**Exercise 4.5.** Let $C \subseteq \mathbb{R}^n$ be a finitely generated cone.

(i) Prove that if $x \notin C$, there exists $\alpha \in C^\circ$ with $\alpha^t x > 0$.
(ii) Prove that
$$(C^\circ)^\circ = C.$$

**Solution 4.5.**    (i) By Theorem 4.12 (Weyl) there exists $A \in \mathbb{R}^{m \times n}$ with $C = \{x \in \mathbb{R}^n \mid Ax \leq 0\}$. Recall that

$$C^\circ = \{\alpha \in \mathbb{R}^n \mid \alpha^t x \leq 0 \text{ for every } x \in C\},$$

from which it follows that $a_1, \ldots, a_m \in C^\circ$ where $a_i$ is the $i$-th row of $A$. If $x \notin C$, one of the inequalities in $Ax \leq 0$ must be violated; this means that there exists an $i$ such that $a_i x > 0$.

(ii) If $x \in C$ then $\alpha^t x \leq 0$ for every $\alpha \in C^\circ$ and thus, $x \in (C^\circ)^\circ$. On the other hand if $x \notin C$, we have just shown that there exists $\alpha \in C^\circ$ such that $\alpha^t x > 0$, implying that $x \notin (C^\circ)^\circ$.

$$\star \quad \star \quad \star$$

**Exercise 4.6.** Let $P = \{x \in \mathbb{R}^d \mid Ax \leq b\} \subseteq \mathbb{R}^d$ be a polyhedron. Prove that if $P$ contains a line $\{v + tu \mid t \in \mathbb{R}\}$ with $u \neq 0$, then $Au = 0$.

**Solution 4.6.** If $P$ contains a line, $tAu \leq b - Av$ for any $t \in \mathbb{R}$. Since $b - Av \geq 0$ is just a fixed vector, this inequality will hold for every $t$ if and only if $Au = 0$.

$$\star \quad \star \quad \star$$

**Exercise 4.7.** Prove that the recession cone of a polyhedron $P = \{x \in \mathbb{R}^n \mid Ax \leq b\}$ is
$$\text{rec}(P) = \{x \in \mathbb{R}^n \mid Ax \leq 0\}.$$

**Solution 4.7.** Suppose that $x \in \text{rec}(P)$. Since $\text{rec}(P)$ is a cone (see Exercise 3.18), $\lambda x \in \text{rec}(P)$ for every $\lambda \geq 0$, thus

$$\lambda Ax \leq b - Av \tag{4.6}$$

for all $\lambda \geq 0$ and $v \in P$. Since $b - Av \geq 0$ is a fixed vector, (4.6) will hold for all $\lambda \geq 0$ if and only if $Ax \leq 0$. On the other hand, let $x \in \mathbb{R}^n$ satisfy $Ax \leq 0$. Accordingly, $A(x + v) \leq Av \leq b$ for all $v \in P$ and $x \in \text{rec}(P)$.

$$\star \quad \star \quad \star$$

**Exercise 4.8.** Use [U.C., (4.13)] to prove that a steady state for a Markov chain with

$$P = \begin{pmatrix} p_{11} & p_{21} \\ p_{12} & p_{22} \end{pmatrix}$$

and $p_{12} + p_{21} > 0$ is

$$x = \left( \frac{p_{21}}{p_{12} + p_{21}}, \frac{p_{12}}{p_{12} + p_{21}} \right).$$

**Solution 4.8.** A vector $x = (x_1, x_2) \in \mathbb{R}^2$ is a steady state if $x = Px$, $x_1 + x_2 = 1$, and $x_1, x_2 \geq 0$. Consequently, we are looking for a non-negative solution to

$$\begin{pmatrix} p_{11} - 1 & p_{21} \\ p_{12} & p_{22} - 1 \\ 1 & 1 \end{pmatrix} x = \begin{pmatrix} 0 \\ 0 \\ 1 \end{pmatrix}. \tag{4.7}$$

Given that the first equation is satisfied, the second will hold as well (remember that the columns of $P$ sum to one). After leaving out the second equation, the remaining system has a unique solution since $p_{12} + p_{21} > 0$. Therefore we find that

$$x = \begin{pmatrix} p_{11} - 1 & p_{21} \\ 1 & 1 \end{pmatrix}^{-1} \begin{pmatrix} 0 \\ 1 \end{pmatrix} = \left( \frac{p_{21}}{p_{12} + p_{21}}, \frac{p_{12}}{p_{12} + p_{21}} \right)^t,$$

which has non-negative entries.

$\star \quad \star \quad \star$

**Exercise 4.9.** Show that a bounded polyhedron is a polytope. Is a polytope a bounded polyhedron?

**Solution 4.9.** Theorem 4.24 gives that

$$P = \text{conv}(\text{ext}(P)) + \text{rec}(P).$$

From Exercise 3.19, we find that $\text{rec}(P) = \{0\}$ since $P$ is bounded and therefore, $P$ is a polytope. (Note that Theorem 4.7 implies $\text{ext}(P) \neq \emptyset$ and Corollary 4.5 that $|\text{ext}(P)| < \infty$.)

Using of Theorem 4.23 (Minkowski and Weyl) with $C = \{0\}$ we find that a polytope is a polyhedron. Furthermore, a polytope is contained in the ball $B(0, r)$ where $r$ is the largest norm among its generators and thus, it is bounded.

$\star \quad \star \quad \star$

**Exercise 4.10.** Give an example of a doubly stochastic $3 \times 3$ matrix with non-zero pairwise different entries. The procedure outlined in [U.C., (4.32)] stops in four steps. Can you give an example of a doubly stochastic $3 \times 3$ matrix, where it stops in five or six steps with suitable choices of permutation matrices along the way?

**Solution 4.10.** The doubly stochastic matrix

$$\frac{1}{50}\begin{pmatrix} 15 & 17 & 18 \\ 16 & 31 & 3 \\ 19 & 2 & 29 \end{pmatrix}$$

has indeed pairwise different entries. Successively subtracting

$$\begin{pmatrix} 1 & 0 & 0 \\ 0 & 1 & 0 \\ 0 & 0 & 1 \end{pmatrix}, \begin{pmatrix} 0 & 1 & 0 \\ 1 & 0 & 0 \\ 0 & 0 & 1 \end{pmatrix}, \begin{pmatrix} 0 & 0 & 1 \\ 0 & 1 & 0 \\ 1 & 0 & 0 \end{pmatrix}, \begin{pmatrix} 0 & 1 & 0 \\ 0 & 0 & 1 \\ 1 & 0 & 0 \end{pmatrix}, \text{ and } \begin{pmatrix} 0 & 0 & 1 \\ 1 & 0 & 0 \\ 0 & 1 & 0 \end{pmatrix}$$

as outlined in [U.C., (4.32)] gives five steps. It is not possible for the procedure to end in six steps. We know by Theorem 4.26 (Birkhoff, von Neumann) that any doubly stochastic matrix is a convex combination of the six permutation matrices. When following the procedure, the first step will make an entry equal to zero and thereby exclude one of the five remaining permutation matrices to be applicable in future steps. As a consequence, the procedure always terminates after at most five steps.

$$\star \quad \star \quad \star$$

**Exercise 4.11.** Let $P_m \subseteq \mathbb{R}^{m \times m}$ be the polyhedron given by

$$
\begin{aligned}
x_{ij} &\geq 0, & \text{for } i, j = 1, \ldots, m \\
x_{i1} + \cdots + x_{im} &\leq 1, & \text{for } i = 1, \ldots, m \\
x_{1j} + \cdots + x_{mj} &\leq 1, & \text{for } j = 1, \ldots, m \\
\sum_{1 \leq i, j \leq m} x_{ij} &\geq m - 1.
\end{aligned}
$$

Prove that $P_m$ is a polytope and that $\dim P_m = m^2$. Construct a natural affine map $f$ from $P_n$ to the Birkhoff polytope $B_{n+1}$. Use $f$ to prove that the dimension of the Birkhoff polytope is

$$\dim B_n = (n-1)^2.$$

**Solution 4.11.** In Exercise 4.9 we proved that a bounded polyhedron is a polytope and since every coordinate in an element of $P_m$ lies between zero and one, $P_m$ is a polytope. Next we let $P'_m$ be the $m \times m$ matrices with entries satisfying

$$x_{ij} > 0 \quad \text{for } i, j = 1, 2, \ldots, m$$
$$x_{i1} + x_{i2} + \cdots + x_{im} < 1 \quad \text{for } i = 1, 2, \ldots, m$$
$$x_{1j} + x_{2j} + \cdots + x_{mj} < 1 \quad \text{for } j = 1, 2, \ldots, m$$
$$\sum_{1 \le i,j \le m} x_{ij} > m - 1.$$

We have $(m - \frac{1}{2})/m^2 \in P'_m$ and that $P'_m$ is an open subset, hence Exercise 2.12 shows $\dim P'_m = m^2$, and the inclusions $P'_m \subseteq P_m \subseteq \mathbb{R}^{m \times m}$ ensure that $\dim P_m = m^2$.

We are now pursuing an affine map $f : \mathbb{R}^{n \times n} \to \mathbb{R}^{(n+1) \times (n+1)}$ such that $f(P_n) = B_{n+1}$. Luckily this can be done. We start by considering the affine maps $g_1, g_2 : \mathbb{R}^{n \times n} \to \mathbb{R}^n$ and $g_3 : \mathbb{R}^{n \times n} \to \mathbb{R}$ given as

$$g_1(x) = \left(1 - \sum_{i=1}^{n} x_{i1}, \ldots, 1 - \sum_{i=1}^{n} x_{in}\right)^t$$

$$g_2(x) = \left(1 - \sum_{j=1}^{n} x_{1j}, \ldots, 1 - \sum_{j=1}^{n} x_{nj}\right)^t$$

$$g_3(x) = \sum_{1 \le i,j \le n} x_{ij} - (n - 1).$$

(Note here that $x$ denotes $n \times n$ matrix.) With these maps we may construct the affine map $f : \mathbb{R}^{n \times n} \to \mathbb{R}^{(n+1) \times (n+1)}$ as

$$f : x \mapsto \begin{pmatrix} x & g_1(x) \\ g_2(x)^t & g_3(x) \end{pmatrix}.$$

The reader is encouraged to go to [U.C., (4.29)-(4.31)] to verify that $f(P_n) = B_{n+1}$.

To prove that $P_n$ and $B_{n+1}$ have the same dimension we argue that $x_1, \ldots, x_d \in P_n$ are affinely independent if and only if $f(x_1), \ldots, f(x_d) \in B_{n+1}$ are affinely independent. Consider the equations

$$\lambda_1 x_1 + \cdots + \lambda_d x_d = 0, \tag{4.8}$$
$$\lambda_1 g_i(x_1) + \cdots + \lambda_d g_i(x_d) = 0 \quad \text{for } i = 1, 2, 3, \tag{4.9}$$
$$\lambda_1 + \cdots + \lambda_d = 0. \tag{4.10}$$

The three equations in (4.9) are redundant since, for $\lambda_1, \ldots, \lambda_d \in \mathbb{R}$ satisfying (4.8) and (4.10),

$$
\begin{aligned}
\lambda_1 g_i(x_1) &+ \cdots + \lambda_d g_i(x_d) \\
&= \lambda_1 \big[ g_i(x_1) - g_i(0) \big] + \cdots + \lambda_d \big[ g_i(x_d) - g_i(0) \big] \\
&= g_i(\lambda_1 x_1 + + \cdots + \lambda_d x_d) - g_i(0) \\
&= 0,
\end{aligned}
$$

where we have used Exercise 2.6 that proves $x \mapsto g_i(x) - g_i(0)$ is linear if $x \mapsto g_i(x)$ is affine. Now Proposition 2.9(3) establishes the equivalence of affine independence in $P_n$ and $B_{n+1}$.

$$\star \quad \star \quad \star$$

**Exercise 4.12.** Prove that a doubly stochastic matrix $A \in B_n$ is a convex combination of at most $n^2 - 2n + 2$ permutation matrices.

**Solution 4.12.** Theorem 4.26 (Birkhoff, von Neumann) gives that if $A \in B_n$, then $A \in \text{conv}(\{E_1, \ldots, E_{n!}\})$ where $E_1, \ldots, E_{n!}$ are the permutation matrices in $\mathbb{R}^{n \times n}$. Corollary 3.15 says that $A$ belongs to the convex hull of an affinely independent subset of the permutation matrices. Since Exercise 4.11 shows $\dim B_n = (n-1)^2$, we conclude that $A$ can be written as at most $(n-1)^2 + 1$ permutation matrices.

$$\star \quad \star \quad \star$$

**Exercise 4.13.** Prove that the graph below (see Example 4.28) does not have a perfect pairing.

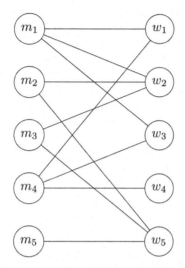

**Solution 4.13.** We have that $m_2$, $m_3$ and $m_5$ only communicate with $w_2$ and $w_5$, thus it is impossible to construct a perfect pairing.

# Chapter 5

# Computations with polyhedra

## 5.1 Introduction

Just as the theory in this chapter the exercises are mainly concerned with two computational tools, namely the double description method and the simplex algorithm. The double description method is about 'describing a set in two ways' as it shows how to convert a half plane representation into a vertex representation and vice versa. Such a tool is very powerful; for example, given a half plane representation it can be hard to answer the question whether it is empty or not, while from a vertex representation it is immediately clear if the set contains an element. The vertex representation cannot be used when considering the simplex algorithm, and for this it will be necessary to convert it into a half plane representation.

Besides computational exercises, some are of theoretical interest and these serve both as a change of pace and to strengthen the conceptual understanding.

The double description method is space consuming to write out in detail in every solution and for this reason, only Solution 5.3 contains a step-by-step solution to a double description exercise. Consequently, the reader should use Solution 5.3 as a computational guideline while the related exercises in question are more result-based and thus, they can preferably be used as such.

**Definition 5.1.** Let $C$ be a convex cone. A non-zero $r \in C$ is called an *extreme ray* in $C$ if $\mathrm{cone}(\{r\})$ is a face of $C$.

With this definition one verifies that $r \in C$ is an extreme ray if and only if

$$r = x + y \quad \text{implies} \quad x, y \in \mathrm{cone}(\{r\}) \tag{5.1}$$

for $x, y \in C$.

**Theorem 5.3.** *Let $A$ be a matrix of full rank $d$ with rows $a_1, \ldots, a_m \in \mathbb{R}^d$. Consider the polyhedral cone*

$$C = \{x \in \mathbb{R}^d \mid Ax \leq 0\}.$$

(1) *A vector $r \in C$ is an extreme ray if and only if $\operatorname{rk} A_r = d - 1$.*

(2) *Two extreme rays $r_1, r_2 \in C$ span a two-dimensional face $\operatorname{cone}(\{r_1, r_2\})$ of $C$ if and only if $\operatorname{rk} A_J = d - 2$, where $J = I(r_1) \cap I(r_2)$.*

(3) *Every element of $C$ is a sum of (finitely many) extreme rays in $C$.*

**Proposition 5.6.** *Let $A$ be an invertible $d \times d$ matrix. Then*

$$C = \{x \in \mathbb{R}^d \mid Ax \leq 0\} = \{Rz \mid z \geq 0\},$$

*where $R = -A^{-1}$. The extreme rays of $C$ are the columns of $R$.*

Note that this theorem gives a way of finding a double description pair $(A, R)$ when $A$ is invertible. This will be serve as the initial step in the double description method.

**Corollary 5.10.** *Suppose that $r_1, \ldots, r_m \in \mathbb{R}^d$ are the extreme rays of*

$$C = \operatorname{cone}(\{r_1, \ldots, r_m\}) = \{x \in \mathbb{R}^d \mid Ax \leq 0\}$$

*for some $n \times d$ matrix $A$. Let $\alpha \in \mathbb{R}^d$. Then*

$$\{r_j \mid \alpha^t r_j \leq 0\} \cup \{(\alpha^t r_i) r_j - (\alpha^t r_j) r_i$$
$$\mid \alpha^t r_i > 0, \ \alpha^t r_j < 0, \ and \ \operatorname{cone}(\{r_i, r_j\}) \ is \ a \ face \ of \ C\}$$

*are the extreme rays of $C \cap \{x \in \mathbb{R}^d \mid \alpha^t x \leq 0\}$.*

This corollary gives the recursive step in the double description method. It shows how to obtain a conic representation of the intersection of a finitely generated conic representation and a half plane. The representation will only contain non-redundant generators.

## 5.2  Exercises and solutions

**Exercise 5.1.** Let $A$ be an $n \times d$ matrix of rank $< d$ and put

$$C = \{x \in \mathbb{R}^d \mid Ax \leq 0\}.$$

(i) Prove that there exists a non-zero vector $z \in \mathbb{R}^d$ with $Az = 0$, $z \in C$ and that such a $z$ is not an extreme ray of $C$.

(ii) If $z \in \mathbb{R}^d$ is non-zero with $Az = 0$ use the identity

$$r = \tfrac{1}{2}(r + z) + \tfrac{1}{2}(r - z)$$

to prove that $C$ cannot have any extreme rays.

**Solution 5.1.**     (i) Since $\operatorname{rk} A < d$, the $d$ columns in $A$ are linearly dependent, so there exists a non-zero vector $z \in \mathbb{R}^d$ such that $Az = 0$. It follows that $-z, 2z \in C$, and since $z = -z + 2z$ and $-z \notin \operatorname{cone}(\{z\})$, (5.1) implies that $z$ is not an extreme ray.

(ii) Suppose that $r \in C$ is an extreme ray. The identity

$$r = \tfrac{1}{2}(r + z) + \tfrac{1}{2}(r - z)$$

implies we can write $r - z = \lambda r$ for some $\lambda \geq 0$ (see (5.1)) implying $z = (1 - \lambda)r$. If $\lambda < 1$ we have that $z$ is an extreme ray, while $\lambda > 1$ implies that $-z$ is an extreme ray. We cannot have $\lambda = 1$, and none of the other scenarios are possible according to (i) resulting in a contradiction.

$$\star \quad \star \quad \star$$

**Exercise 5.2.** Prove Proposition 5.6 using Theorem 5.3(1) and the identity

$$AA^{-1} = I_d,$$

where $I_d$ is the $d \times d$ identity matrix.

**Solution 5.2.** Let $x \in \mathbb{R}^d$ and suppose that $Ax \leq 0$. Setting $z = -Ax$ it follows that $x = -A^{-1}z = Rz$ and $z \geq 0$. On the other hand, if $x \in \mathbb{R}^d$ satisfies $x = Rz$ for some $z \geq 0$ we get that

$$Ax = A(Rz) = -AA^{-1}z = -z \leq 0.$$

Let $r_i$ denote the $i$-th column of $R$. After noting that $C = \{Rz \mid z \geq 0\} = \operatorname{cone}(\{r_1, \ldots, r_d\})$, Lemma 5.2 shows that the set of extreme rays of $C$ (up to a positive scalar) is included in $\{r_1, \ldots, r_d\}$. For any $i$, observe that $I(r_i) = \{1, \ldots, i-1, i+1, \ldots, d\}$, since $Ar_i = -e_i$. (Here $e_i$ is the $i$-th canonical basis vector.) Thus, $r_i$ is an extreme ray according to Theorem 5.3(1).

$$\star \quad \star \quad \star$$

**Exercise 5.3.** Use the double description method to verify [U.C., (5.5)].

**Solution 5.3.** We consider the polyhedral cone $\hat{P} = \{x \in \mathbb{R}^4 \mid Ax \le 0\}$ where

$$
A = \begin{pmatrix} — a_1 — \\ — a_2 — \\ — a_3 — \\ — a_4 — \\ — a_5 — \\ — a_6 — \\ — a_7 — \end{pmatrix} = \begin{pmatrix} -1 & 0 & 0 & 1 \\ -1 & 0 & -1 & 2 \\ 0 & -1 & 0 & 1 \\ 0 & -1 & -1 & 2 \\ 1 & 1 & -1 & -4 \\ 1 & 1 & 0 & -5 \\ 0 & 0 & 0 & -1 \end{pmatrix}.
$$

First note that

$$
\hat{P} = \{x \in \mathbb{R}^4 \mid A_{\{1,2,3,5\}} x \le 0\} \cap \bigcap_{i=4,6,7} \{x \in \mathbb{R}^4 \mid a_i x \le 0\}.
$$

An application of Proposition 5.6 gives

$$
\{x \in \mathbb{R}^4 \mid A_{\{1,2,3,5\}} x \le 0\} = \operatorname{cone}\left(\left\{\begin{pmatrix} 5 \\ 2 \\ -1 \\ 2 \end{pmatrix}, \begin{pmatrix} -1 \\ -1 \\ 2 \\ -1 \end{pmatrix}, \begin{pmatrix} 1 \\ 4 \\ 1 \\ 1 \end{pmatrix}, \begin{pmatrix} 1 \\ 1 \\ 1 \\ 1 \end{pmatrix}\right\}\right). \qquad (5.2)
$$

For instance, to write the intersection of (5.2) and $\{x \in \mathbb{R}^4 \mid a_4 x \le 0\}$ as a cone we apply Corollary 5.10. Thus, we start by letting $r_i$ be the $i$-th generator in (5.2) and we compute $a_4 r_1 = 3$, $a_4 r_2 = -3$, $a_4 r_3 = -3$, and $a_4 r_4 = 0$. Since $a_4 r_1 > 0$, $a_4 r_2 < 0$, and $a_4 r_3 < 0$, we should check if $\operatorname{cone}(\{r_1, r_2\})$ and $\operatorname{cone}(\{r_1, r_3\})$ are faces of $\{x \in \mathbb{R}^4 \mid A_{\{1,2,3,5\}} x \le 0\}$. However, this is always the case in the first double description step by the structure of $I(r_i)$ (cf. Solution 5.2 and Theorem 5.3(2)). We compute

$$
(a_4 r_1) r_2 - (a_4 r_2) r_1 = 3r_2 + 3r_1 \quad \text{and} \quad (a_4 r_1) r_3 - (a_4 r_3) r_1 = 3r_3 + 3r_1,
$$

and since $\{r_j \mid a_4 r_j \le 0\} = \{r_2, r_3, r_4\}$ we conclude that

$$
\{x \in \mathbb{R}^4 \mid A_{\{1,2,3,5\}} x \le 0\} \cap \{x \in \mathbb{R}^4 \mid a_4 x \le 0\}
$$
$$
= \operatorname{cone}(\{r_2, r_3, r_4, r_2 + r_1, r_3 + r_1\})
$$
$$
= \operatorname{cone}\left(\left\{\begin{pmatrix} -1 \\ -1 \\ 2 \\ -1 \end{pmatrix}, \begin{pmatrix} 1 \\ 4 \\ 1 \\ 1 \end{pmatrix}, \begin{pmatrix} 1 \\ 1 \\ 1 \\ 1 \end{pmatrix}, \begin{pmatrix} 4 \\ 1 \\ 1 \\ 1 \end{pmatrix}, \begin{pmatrix} 2 \\ 2 \\ 0 \\ 1 \end{pmatrix}\right\}\right), \qquad (5.3)
$$

where all rays are extreme.

As the next step we start by renaming $r_i$ as the $i$-th generator in (5.3). To intersect the cone with $\{x \in \mathbb{R}^4 \mid a_6 x \le 0\}$ we compute $a_6 r_1 = 3$, $a_6 r_2 = 0$, $a_6 r_3 = -3$, $a_6 r_4 = 0$, and $a_6 r_5 = -1$. Consequently, we have to check if cone($\{r_1, r_2\}$) and cone($\{r_1, r_3\}$) are faces of $\{x \in \mathbb{R}^4 \mid A_{\{1,2,3,4,5\}}x \leq 0\}$. We find $I(r_1) \cap I(r_2) = \{5\}$ and $I(r_1) \cap I(r_3) = \{1, 3\}$ and hence, we have by Theorem 5.3(2) that only cone($\{r_1, r_3\}$) is a face. As a consequence,

$$\{x \in \mathbb{R}^4 \mid A_{\{1,2,3,4,5\}}x \leq 0\} \cap \{x \in \mathbb{R}^4 \mid a_6 x \leq 0\}$$
$$= \text{cone}(\{r_2, r_3, r_4, r_5, r_1 + r_3\})$$
$$= \text{cone}\left(\left\{\begin{pmatrix}1\\4\\1\\1\end{pmatrix}, \begin{pmatrix}1\\1\\1\\1\end{pmatrix}, \begin{pmatrix}4\\1\\1\\1\end{pmatrix}, \begin{pmatrix}2\\2\\0\\1\end{pmatrix}, \begin{pmatrix}0\\0\\1\\0\end{pmatrix}\right\}\right). \tag{5.4}$$

It appears that (5.4) coincides with the cone given [U.C., (5.5)], and since $\{x \in \mathbb{R}^4 \mid a_7 x \leq 0\}$ is a redundant restriction, the result is verified.

$$\star \quad \star \quad \star$$

**Exercise 5.4.** Use the double description method to verify [U.C., (5.7)].

**Solution 5.4.** Using Proposition 5.6 we find

$$\left\{x \in \mathbb{R}^4 \left| \begin{pmatrix}1&1&1&1\\2&1&1&1\\1&2&1&1\\2&2&3&1\end{pmatrix} x \leq 0\right.\right\} = \text{cone}\left(\left\{\begin{pmatrix}2\\2\\-1\\-5\end{pmatrix}, \begin{pmatrix}-2\\0\\1\\1\end{pmatrix}, \begin{pmatrix}0\\-2\\1\\1\end{pmatrix}, \begin{pmatrix}0\\0\\-1\\1\end{pmatrix}\right\}\right).$$

The procedure for intersecting this set with $\{x \in \mathbb{R}^4 \mid (1, 1, 5, 1)x \leq 0\}$ is the same as in Exercise 5.3, and this verifies [U.C., (5.7)].

$$\star \quad \star \quad \star$$

**Exercise 5.5.** Check if the inequalities

$$\begin{aligned}
2x - 3y + z &\leq -2\\
x + 3y + z &\leq -3\\
-2x - 3y + z &\leq -2\\
-x - 3y - 3z &\leq 1\\
-2x - y + 3z &\leq 3
\end{aligned}$$

have a solution $x, y, z \in \mathbb{R}$.

**Solution 5.5.** The existence of a solution turns into the question whether

$$\begin{pmatrix}x\\y\\z\\1\end{pmatrix} \in \left\{v \in \mathbb{R}^4 \left| \begin{pmatrix}2&-3&1&2\\1&3&1&3\\-2&-3&1&2\\-1&-3&-3&-1\\-2&-1&3&-3\end{pmatrix} v \leq 0\right.\right\}. \tag{5.5}$$

An application of Proposition 5.6 immediately yields

$$\left\{ v \in \mathbb{R}^4 \,\middle|\, \begin{pmatrix} 2 & -3 & 1 & 2 \\ 1 & 3 & 1 & 3 \\ -2 & -3 & 1 & 2 \\ -1 & -3 & -3 & -1 \end{pmatrix} v \le 0 \right\} = \text{cone}\left( \left\{ \begin{pmatrix} -21 \\ 11 \\ -3 \\ -3 \end{pmatrix}, \begin{pmatrix} 0 \\ -10 \\ 18 \\ -24 \end{pmatrix}, \begin{pmatrix} 21 \\ 5 \\ -9 \\ -9 \end{pmatrix}, \begin{pmatrix} 0 \\ 2 \\ 30 \\ -12 \end{pmatrix} \right\} \right).$$

No vector has a positive fourth coordinate in this set and thus, (5.5) cannot hold.

$$\star \quad \star \quad \star$$

**Exercise 5.6.** Let $P$ be the set of points $(x, y, z) \in \mathbb{R}^3$ satisfying

$$
\begin{aligned}
-x - 2y + 3z &\le -2 \\
-3x - y + 2z &\le -1 \\
x &\ge 0 \\
y &\ge 0 \\
z &\ge 0.
\end{aligned}
$$

(i) Compute the extreme points of $P$.

(ii) Compute the extreme rays in the recession cone of $P$.

(iii) Write $P = C + Q$ as in [U.C., (5.1)].

**Solution 5.6.** (i) We use the method explained in relation to Corollary 4.5 when the setup is $P = \{v \in \mathbb{R}^3 \mid Av \le b\}$ with

$$A = \begin{pmatrix} -1 & -2 & 3 \\ -3 & -1 & 2 \\ -1 & 0 & 0 \\ 0 & -1 & 0 \\ 0 & 0 & -1 \end{pmatrix} \quad \text{and} \quad b = \begin{pmatrix} -2 \\ -1 \\ 0 \\ 0 \\ 0 \end{pmatrix}.$$

By choosing $I_1 = \{1, 2, 3\}$ we get that rk $A_{I_1} = 3$ and $A_{I_1}^{-1} b_{I_1} = (0, 1, 0)^t$. Since $(0, 1, 0) \in P$ we have found an extreme point. If we now choose $I_2 = \{1, 2, 4\}$, then rk $A_{I_2} = 3$ but $A_{I_2}^{-1} b_{I_2} = (-1/7, 0, -5/7)^t \notin P$, and therefore the point is not extreme. Continuing this way yields $\text{ext}(P) = \{(0, 1, 0), (2, 0, 0)\}$.

(ii) According to Exercise 4.7 we have that $\text{rec}(P) = \{v \in \mathbb{R}^3 \mid Av \le 0\}$. To find the extreme rays of $\text{rec}(P)$ we use the same procedure as in Example 5.5. By choosing $J = \{1, 2\}$ we find that rk $A_J = 2$ and $\{v \in \mathbb{R}^3 \mid A_J v = 0\} = \mathbb{R}z$ with $z = (1, 5, 7)$. This vector belongs to $\text{rec}(P)$ if and only if $z \ge 0$, thus we have that $r_1 = (1, 7, 5)$ is an extreme ray. If we take $J = \{1, 3\}$ the only solution in $\text{rec}(P)$ is $(0, 0, 0)$, but an extreme ray is by definition a non-zero vector. After traversing the other possible choices of $J \subseteq \{1, 2, 3, 4, 5\}$ with $|J| = 2$ we find that only $J = \{2, 3\}$ leads to another extreme ray, namely $r_2 = (0, 2, 1)$. Thus, up to multiplication by a scalar, $r_1$ and $r_2$ are the extreme rays of $\mathrm{rec}(P)$ meaning that $\mathrm{rec}(P) = \mathrm{cone}(\{r_1, r_2\})$.

(iii) It follows directly from Theorem 4.24 that $P = Q + C$, where

$$Q = \mathrm{conv}\left(\left\{\begin{pmatrix} 0 \\ 1 \\ 0 \end{pmatrix}, \begin{pmatrix} 2 \\ 0 \\ 0 \end{pmatrix}\right\}\right) \quad \text{and} \quad C = \mathrm{rec}(P) = \mathrm{cone}\left(\left\{\begin{pmatrix} 1 \\ 7 \\ 5 \end{pmatrix}, \begin{pmatrix} 0 \\ 2 \\ 1 \end{pmatrix}\right\}\right).$$

Remark how this procedure converts a half space representation to a vertex representation. For this reason it serves as an alternative to the double description method (see §5.3.1), and one could as well solve the exercise using this method.

$$\star \quad \star \quad \star$$

**Exercise 5.7.** Let $P$ denote the set of $(x_1, x_2, x_3, x_4) \in \mathbb{R}^4$ satisfying

$$
\begin{aligned}
x_1 + x_2 + x_3 + x_4 &\geq 1 \\
-3x_1 + x_2 + x_3 + x_4 &\leq 1 \\
x_1 - 3x_2 + x_3 + x_4 &\leq 1 \\
x_1 + x_2 - 3x_3 + x_4 &\leq 1 \\
x_1 + x_2 + x_3 - 3x_4 &\leq 1.
\end{aligned}
$$

Express $P$ as in [U.C., (5.1)]. Is $P$ bounded? Does $P$ contain a 4-simplex?

**Solution 5.7.** We use the procedure as in §5.3.1. Define the polyhedral cone

$$\hat{P} = \left\{ (x, z) \in \mathbb{R}^5 \mid x \in \mathbb{R}^4, \ z \in \mathbb{R}, \ \hat{A}\begin{pmatrix} x \\ z \end{pmatrix} \leq 0 \right\}$$

where

$$\hat{A} = \begin{pmatrix} -1 & -1 & -1 & -1 & 1 \\ -3 & 1 & 1 & 1 & -1 \\ 1 & -3 & 1 & 1 & -1 \\ 1 & 1 & -3 & 1 & -1 \\ 1 & 1 & 1 & -3 & -1 \\ 0 & 0 & 0 & 0 & -1 \end{pmatrix}.$$

Using Proposition 5.6 we find

$$\hat{P} = \mathrm{cone}\left(\left\{\begin{pmatrix} 1 \\ 1 \\ 1 \\ 1 \\ 0 \end{pmatrix}, \begin{pmatrix} 1 \\ 0 \\ 0 \\ 0 \\ 1 \end{pmatrix}, \begin{pmatrix} 0 \\ 1 \\ 0 \\ 0 \\ 1 \end{pmatrix}, \begin{pmatrix} 0 \\ 0 \\ 1 \\ 0 \\ 1 \end{pmatrix}, \begin{pmatrix} 0 \\ 0 \\ 0 \\ 1 \\ 1 \end{pmatrix}\right\}\right) \cap \{(x, z) \in \mathbb{R}^5 \mid z \geq 0\}.$$

After noting that the restriction $z \geq 0$ is redundant we conclude that

$$P = \text{cone}\left(\left\{\begin{pmatrix}1\\1\\1\\1\end{pmatrix}\right\}\right) + \text{conv}\left(\left\{\begin{pmatrix}1\\0\\0\\0\end{pmatrix}, \begin{pmatrix}0\\1\\0\\0\end{pmatrix}, \begin{pmatrix}0\\0\\1\\0\end{pmatrix}, \begin{pmatrix}0\\0\\0\\1\end{pmatrix}\right\}\right).$$

In particular, $P$ is unbounded since the cone is non-zero. Furthermore, we note that

$$\text{conv}\left(\left\{\begin{pmatrix}1\\1\\1\\1\end{pmatrix}, \begin{pmatrix}1\\0\\0\\0\end{pmatrix}, \begin{pmatrix}0\\1\\0\\0\end{pmatrix}, \begin{pmatrix}0\\0\\1\\0\end{pmatrix}, \begin{pmatrix}0\\0\\0\\1\end{pmatrix}\right\}\right) \subseteq P,$$

which is a 4-simplex.

★    ★    ★

**Exercise 5.8.** Write down a minimal set of inequalities in $x$, $y$ and $z$ describing if

$$\begin{pmatrix}x\\y\\z\end{pmatrix}$$

is in the convex hull of

$$\begin{pmatrix}1\\1\\0\end{pmatrix}, \quad \begin{pmatrix}1\\-1\\0\end{pmatrix}, \quad \begin{pmatrix}-1\\-1\\0\end{pmatrix}, \quad \begin{pmatrix}-1\\1\\0\end{pmatrix}, \quad \begin{pmatrix}0\\0\\1\end{pmatrix} \quad \text{and} \quad \begin{pmatrix}0\\0\\-1\end{pmatrix}$$

in $\mathbb{R}^3$. A minimal inequality corresponds to a facet. Show that the convex hull of five points not all in the same affine hyperplane of $\mathbb{R}^3$ has at most six facets.

**Solution 5.8.** We are given the polytope

$$P = \text{conv}\left(\left\{\begin{pmatrix}1\\1\\0\end{pmatrix}, \begin{pmatrix}1\\-1\\0\end{pmatrix}, \begin{pmatrix}-1\\-1\\0\end{pmatrix}, \begin{pmatrix}-1\\1\\0\end{pmatrix}, \begin{pmatrix}0\\0\\1\end{pmatrix}, \begin{pmatrix}0\\0\\-1\end{pmatrix}\right\}\right)$$

and are asked to find a half space representation of $P$. This problem is encountered as a part of the double description method in §5.3.2. Define

$$R = \begin{pmatrix} 1 & 1 & -1 & -1 & 0 & 0 \\ 1 & -1 & -1 & 1 & 0 & 0 \\ 0 & 0 & 0 & 0 & 1 & -1 \\ 1 & 1 & 1 & 1 & 1 & 1 \end{pmatrix}.$$

The task is to find $\hat{A}^t$ such that $(R^t, \hat{A}^t)$ is a double description pair. Let $\hat{P} = \{v \in \mathbb{R}^4 \mid R^t v \leq 0\}$. We are now in a familiar setup, and we can apply the double description method as in Exercise 5.3. This leads to

$$\hat{P} = \text{cone}\left(\left\{\begin{pmatrix}-1\\0\\-1\\-1\end{pmatrix}, \begin{pmatrix}-1\\0\\1\\-1\end{pmatrix}, \begin{pmatrix}0\\1\\-1\\-1\end{pmatrix}, \begin{pmatrix}0\\1\\1\\-1\end{pmatrix}, \begin{pmatrix}0\\-1\\-1\\-1\end{pmatrix}, \begin{pmatrix}0\\-1\\1\\-1\end{pmatrix}, \begin{pmatrix}1\\0\\-1\\-1\end{pmatrix}, \begin{pmatrix}1\\0\\1\\-1\end{pmatrix}\right\}\right).$$

By defining $\hat{A}$ to be the $8 \times 4$-matrix having the generators in the cone above as its rows, we get the relation

$$\{v \in \mathbb{R}^4 \mid R^t v \leq 0\} = \{\hat{A}^t z \mid z \geq 0\}$$

or in words, $(R^t, \hat{A}^t)$ is a double description pair. We conclude that $(\hat{A}, R)$ is also a double description pair by Lemma 4.21, which means that $(x, y, z) \in P$ if and only if $\hat{A}(x, y, z, 1)^t \leq 0$ that is, if and only if $(x, y, z)$ satisfies that

$$\begin{array}{rcl}
-x & - z \leq 1 \\
-x & + z \leq 1 \\
y & - z \leq 1 \\
y & + z \leq 1 \\
-y & - z \leq 1 \\
-y & + z \leq 1 \\
x & - z \leq 1 \\
x & + z \leq 1.
\end{array}$$

To show the last part of the exercise let $v_1, v_2, v_3, v_4, v_5 \in \mathbb{R}^3$ and $P = \text{conv}(\{v_1, v_2, v_3, v_4, v_5\})$. The strategy is to carry out the same steps as above to show that $\hat{A}$ has at most six rows (each row is a minimal inequality for $P$, which in turn corresponds to a facet). Thus, define $\hat{P} = \{u \in \mathbb{R}^4 \mid R^t u \leq 0\}$ and

$$R = \begin{pmatrix} v_1 & \cdots & v_5 \\ 1 & \cdots & 1 \end{pmatrix}.$$

Note that four of the points in $\{v_1, v_2, v_3, v_4, v_5\}$ are affinely independent since they are not contained in the same hyperplane. Consequently, we can assume that

$$\begin{pmatrix} v_1 & \cdots & v_4 \\ 1 & \cdots & 1 \end{pmatrix}$$

is invertible, which means that we can apply Proposition 5.6 to obtain vectors $r_1, r_2, r_3, r_4 \in \mathbb{R}^4$ with

$$\hat{P} = \text{cone}(\{r_1, r_2, r_3, r_4\}) \cap \{u \in \mathbb{R}^4 \mid (v_5, 1)u \leq 0\}. \tag{5.6}$$

We are able to write (5.6) as a finitely generated cone cf. Corollary 5.10, from which it can be checked that this generating set (of extreme rays) at most contains six vectors. For instance, if $(v_5^t, 1)r_i > 0$ for $i = 1, 2$ and

$(v_5^t, 1)r_i < 0$ for $i = 3, 4$, the generators would be $r_3$, $r_4$ and a subset of $M = \{((v_5, 1)r_i)r_j - ((v_5, 1)r_j)r_i \mid i = 1, 2, \ j = 3, 4\}$ where $|M| = 4$.

$$\star \quad \star \quad \star$$

**Exercise 5.9.** Compute $\mathrm{rec}(P)$, where $P$ is the set of solutions to

$$
\begin{aligned}
-x - \ y + \ z &\le 1 \\
-x + 2y - \ z &\le 2 \\
x - 2y + \ z &\le 1 \\
5x + 5y - 7z &\le 5
\end{aligned}
$$

in $\mathbb{R}^3$.

**Solution 5.9.** By Exercise 4.7, $\mathrm{rec}(P) = \{x \in \mathbb{R}^3 \mid Ax \le 0\}$ with

$$
A = \begin{pmatrix} -1 & -1 & 1 \\ -1 & 2 & -1 \\ 1 & -2 & 1 \\ 5 & 5 & -7 \end{pmatrix}.
$$

The conic representation is given as

$$
\mathrm{rec}(P) = \mathrm{cone}\left(\left\{\begin{pmatrix} 3 \\ 4 \\ 5 \end{pmatrix}, \begin{pmatrix} 1 \\ 2 \\ 3 \end{pmatrix}\right\}\right),
$$

which may be verified using the double description method.

$$\star \quad \star \quad \star$$

**Exercise 5.10.**

(i) Let $P$ denote the set of $(x, y, z) \in \mathbb{R}^3$ satisfying

$$
\begin{aligned}
x + y + z &\le 1 \\
x + y - z &\le 1 \\
x - y - z &\le 1 \\
-x + y - z &\le 1 \\
x - y + z &\le 1 \\
-x + y + z &\le 1 \\
-x - y + z &\le 1 \\
-x - y - z &\le 1.
\end{aligned}
$$

Show that $P$ is the convex hull of 6 points.

(ii) Show that $P$ has 12 one-dimensional faces (edges).

**Solution 5.10.** (i) In line with earlier exercises (for instance, Exercise 5.5) we consider the polyhedral cone

$$
\hat{P} = \left\{ (x,y,z,w) \in \mathbb{R}^4 \,\middle|\, \begin{pmatrix} 1 & 1 & 1 & -1 \\ 1 & 1 & -1 & -1 \\ 1 & -1 & -1 & -1 \\ -1 & 1 & -1 & -1 \\ 1 & -1 & 1 & -1 \\ -1 & 1 & 1 & -1 \\ -1 & -1 & 1 & -1 \\ -1 & -1 & -1 & -1 \end{pmatrix} \begin{pmatrix} x \\ y \\ z \\ w \end{pmatrix} \le 0 \right\}
$$

and apply the double description method to write the set as a finitely generated cone

$$
\hat{P} = \operatorname{cone}\left( \left\{ \begin{pmatrix} 1 \\ 0 \\ 0 \\ 1 \end{pmatrix}, \begin{pmatrix} 0 \\ 1 \\ 0 \\ 1 \end{pmatrix}, \begin{pmatrix} 0 \\ 0 \\ 1 \\ 1 \end{pmatrix}, \begin{pmatrix} -1 \\ 0 \\ 0 \\ 1 \end{pmatrix}, \begin{pmatrix} 0 \\ -1 \\ 0 \\ 1 \end{pmatrix}, \begin{pmatrix} 0 \\ 0 \\ -1 \\ 1 \end{pmatrix} \right\} \right).
$$

From this we obtain

$$
P = \operatorname{conv}\left( \left\{ \begin{pmatrix} 1 \\ 0 \\ 0 \end{pmatrix}, \begin{pmatrix} 0 \\ 1 \\ 0 \end{pmatrix}, \begin{pmatrix} 0 \\ 0 \\ 1 \end{pmatrix}, \begin{pmatrix} -1 \\ 0 \\ 0 \end{pmatrix}, \begin{pmatrix} 0 \\ -1 \\ 0 \end{pmatrix}, \begin{pmatrix} 0 \\ 0 \\ -1 \end{pmatrix} \right\} \right), \qquad (5.7)
$$

which shows that $P$ is the convex hull of six points.

(ii) Note that faces of polyhedra are all exposed, and in the special case of a polytope this property ensures that any face is given as the convex hull of some of its generators. As a consequence, when searching for edges we only need to check if $\operatorname{conv}(\{v_1, v_2\})$ is an edge of $P$ for combinations of $v_1$ and $v_2$ in the generating set of (5.7). Since

$$
\begin{pmatrix} 0 \\ 0 \\ 0 \end{pmatrix} = \tfrac{1}{2}\begin{pmatrix} 1 \\ 0 \\ 0 \end{pmatrix} + \tfrac{1}{2}\begin{pmatrix} -1 \\ 0 \\ 0 \end{pmatrix} = \tfrac{1}{2}\begin{pmatrix} 0 \\ 1 \\ 0 \end{pmatrix} + \tfrac{1}{2}\begin{pmatrix} 0 \\ -1 \\ 0 \end{pmatrix} = \tfrac{1}{2}\begin{pmatrix} 0 \\ 0 \\ 1 \end{pmatrix} + \tfrac{1}{2}\begin{pmatrix} 0 \\ 0 \\ -1 \end{pmatrix},
$$

we see that $(0,0,0)$ is in the convex hull of three pairs of the generators, and therefore none of these can be associated with an edge. By Exercise 4.3 we know that the convex hull of two vertices $v_1, v_2 \in P$ is an edge if $\operatorname{rk} A_z = 2$ where $z = (v_1 + v_2)/2$. Here $A$ is the $8 \times 3$ matrix that corresponds to the representation $P = \{v \in \mathbb{R}^3 \mid Av \le b\}$. For instance, checking $v_1 = (1,0,0)$ and $v_2 = (0,1,0)$ gives $z = (1/2, 1/2, 0)$ and

$$
A_z = \begin{pmatrix} 1 & 1 & 1 \\ 1 & 1 & -1 \end{pmatrix}.
$$

This shows rk $A_z = 2$ and we conclude that conv($\{v_1, v_2\}$) is an edge of $P$. By running through the remaining 11 combinations of vertices in (5.7) we verify that they are all edges of $P$.

$$\star \quad \star \quad \star$$

**Exercise 5.11.** The ball centered at $v \in \mathbb{R}^n$ with radius $r \geq 0$ is the subset

$$B(v, r) := \{x \in \mathbb{R}^n \mid |x - v|^2 \leq r^2\}.$$

Let $H^- = \{x \in \mathbb{R}^n \mid \alpha^t x \leq \beta\}$ be a given affine half space.

(i) Prove that

$$B(v, r) = \{v + \lambda u \mid 0 \leq \lambda \leq r, |u| = 1\}.$$

(ii) Prove that $B(v, r) \subseteq H^-$ if and only if

$$\alpha^t v + |\alpha| r \leq \beta.$$

**Solution 5.11.**     (i) Assume $x \neq v$. If $x \in B(v, r)$, let

$$u = \tfrac{1}{|x-v|}(x - v) \quad \text{and} \quad \lambda = |x - v|.$$

Then $x = v + \lambda u$ with $0 \leq \lambda \leq r$ and $|u| = 1$. On the other hand, if $x = v + \lambda u$ with $0 \leq \lambda \leq r$ and $|u| = 1$ we have $|x - v| = \lambda \leq r$, and this shows that $x \in B(v, r)$.

(ii) Suppose that $B(v, r) \subseteq H^-$. This implies $\alpha^t(v + ru) \leq \beta$ for any $u \in \mathbb{R}^n$ with $|u| = 1$ by (i). Choosing $u = |\alpha|^{-1}\alpha$ shows $\alpha^t v + |\alpha| r \leq \beta$. The other implication follows by Lemma A.1 (Cauchy-Schwarz).

$$\star \quad \star \quad \star$$

**Exercise 5.12.** It also seems that there exists a circle of minimal radius containing the polygon in Example 5.15. Compute the center and radius of this circle.

**Solution 5.12.** Consider a general convex hull conv($\{x_1, \ldots, x_m\}$) with $x_1, \ldots, x_m \in \mathbb{R}^n$. Let $B(v, r)$ be a ball with center $v$ and radius $r$. By convexity of $B(v, r)$ we have that $x_1, \ldots, x_m \in B(v, r)$ if and only if conv($\{x_1, \ldots, x_m\}$) $\subseteq B(v, r)$. This implies that fining the smallest ball containing the convex hull is equivalent of solving

$$\min\{r \mid |x_1 - v| \leq r, \ldots, |x_m - v| \leq r\}. \tag{5.8}$$

Unfortunately this is not a linear program in any dimension greater than one, and a general way of solving such an optimization problem may involve the machinery of Chapter 10. Instead our solution will be based on case specific arguments. Start by writing

$$m = \max\{|x_i - x_j| \mid 1 \leq i, j \leq n\}$$

and let $i^*$ and $j^*$ be the indices where the maximum is attained. Then we have a lower bound on (5.8) equal to $m/2$, since any ball with a radius lower than this cannot contain both $x_{i^*}$ and $x_{j^*}$. On the other hand, if we consider the radius $r = m/2$ and let the center be $v = (x_{i^*} + x_{j^*})/2$, we will be certain that both $x_{i^*}$ and $x_{j^*}$ are in the ball. In addition, if the other $n - 2$ other elements are in the ball we have solved (5.8), since we have a feasible pair $(v, r)$ with $r$ equal to the lower bound.

The strategy outlined above is exactly how we will find a ball with minimum radius containing the polygon

$$P = \mathrm{conv}\left(\left\{\begin{pmatrix}1\\1\end{pmatrix}, \begin{pmatrix}2\\0\end{pmatrix}, \begin{pmatrix}4\\2\end{pmatrix}, \begin{pmatrix}2\\3\end{pmatrix}\right\}\right)$$

from Example 5.15. With notation as above we find

$$m = |(1,1) - (4,2)| = \sqrt{10}$$
$$v = \tfrac{1}{2}(1,1) + \tfrac{1}{2}(4,2) = (\tfrac{5}{2}, \tfrac{3}{2}).$$

Then we calculate

$$|(2,0) - v| = \tfrac{1}{2}\sqrt{10} = \tfrac{1}{2}m$$
$$|(2,3) - v| = \tfrac{1}{2}\sqrt{10} = \tfrac{1}{2}m$$

and conclude that the smallest ball containing $P$ must be

$$B\left((\tfrac{5}{2}, \tfrac{3}{2}), \tfrac{1}{2}\sqrt{10}\right)$$

(see Figure 5.1).

⋆ ⋆ ⋆

**Exercise 5.13.** It may happen that a system

$$a_{11}x_1 + \cdots + a_{1n}x_n = b_1$$
$$\vdots$$
$$a_{m1}x_1 + \cdots + a_{mn}x_n = b_m$$

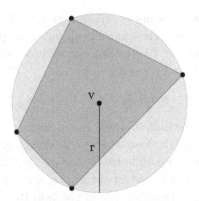

**Figure 5.1:** The circle $B((\frac{5}{2}, \frac{3}{2}), \frac{1}{2}\sqrt{10})$ and the polygon $P$.

of linear equations does not have a solution. In practice one is interested in an approximate "solution", such that the maximal absolute error

$$\max \{|b_1 - (a_{11}x_1 + \cdots + a_{1n}x_n)|, \ldots, |b_m - (a_{m1}x_1 + \cdots + a_{mn}x_n)|\}$$

is minimal. Give a suggestion as to how this problem can be solved.

**Solution 5.13.** Minimizing the maximal absolute error is equivalent of finding a minimal $d \geq 0$ such that $|b_i - (a_{i1}x_1 + \cdots + a_{in}x_n)| \leq d$ for $i = 1, \ldots, m$. Writing each of these restrictions as two inequalities, we get the linear program in compact form as

$$\min \{d \mid Ax - b \leq d\iota, b - Ax \leq d\iota\}$$

with natural definitions of $A$ and $b$, and where $\iota$ is an $m$-dimensional vector of ones. One can now proceed and solve the problem using the simplex algorithm or any other linear programming method. (For more intuition about the problem, see Example 5.16.)

$$\star \quad \star \quad \star$$

**Exercise 5.14.** Let $x, y \in P \subseteq \mathbb{R}^n$ be vertices in a polyhedron $P$. Prove that $y - x$ is an extreme ray in $C_x$ if $\text{conv}\{x, y\}$ is an edge in $P$. Let $d \in C_x$ be an extreme ray. Prove that $\{x + \lambda d \mid \lambda \geq 0\} \cap P$ is an edge of $P$. Does an extreme ray in $C_x$ necessarily lead to a neighboring vertex?

**Solution 5.14.** Recall that $C_x = \{d \in \mathbb{R}^n \mid A_x d \leq 0\}$. Since $\text{conv}(\{x, y\})$ is an edge in $P$ it follows by Proposition 4.3 that $\text{rk} A_z = d - 1$ for some

$z \in \text{conv}(\{x, y\})$. Combining this with the fact that $x$ and $y$ are vertices in $P$ gives that $\text{rk}\, A_{I(x) \cap I(y)} = d - 1$, and since $(A_x(y - x))_i = (A_x y - b_x)_i$ we observe that $(A_x y)_i = (b_x)_i$ if and only if $(A_x(x - y))_i = 0$. Now Theorem 5.3(1) shows that $y - x$ is an extreme ray of $C_x$.

To prove that $\{x + \lambda d \mid \lambda \geq 0\} \cap P$ is an edge in $P$ note that

$$(A_x d)_i = 0 \quad \text{if and only if} \quad (A_x(x + \lambda d))_i = (b_x)_i \qquad (5.9)$$

when $\lambda > 0$. By fixing $\lambda_0 > 0$ sufficiently small we obtain $I(x + \lambda_0 d) \subseteq I(x)$. Then since $d$ is an extreme ray of $C_x$, (5.9) gives that $\text{rk}\, A_{x + \lambda_0 d} = \text{rk}\, A_{I(x) \cap I(x + \lambda_0 d)} = d - 1$, and this means that

$$P_{I(x + \lambda_0 d)} = \{v \in P \mid A_{x + \lambda_0 d} v = b_{x + \lambda_0 d}\}$$

is an edge of $P$ by Proposition 4.3. It follows from Proposition 2.6 that

$$\{v \in \mathbb{R}^n \mid A_{x + \lambda_0 d} v = b_{x + \lambda_0 d}\} = \{x + \lambda d \mid \lambda \in \mathbb{R}\}.$$

For $x + \lambda d \in P$ we must have that $\lambda \geq 0$, since (necessarily) there exists $i$ such that $(Ax)_i = b_i$ and $(Ad)_i < 0$. Therefore we find $P_{I(x + \lambda_0 d)} = \{x + \lambda d \mid \lambda \geq 0\} \cap P$.

An extreme ray in $C_x$ does not always lead to a neighboring vertex. A counterexample is the set

$$P = \{(x, y) \in \mathbb{R}^2 \mid 0 \leq x \leq 2, 0 \leq y\}.$$

Here $(0, 0)$ is a vertex, and $(1, 0)$ and $(0, 1)$ are extreme rays of $C_{(0,0)}$, but $(0, 1)$ does not lead to the other vertex, $(2, 0)$.

$$\star \quad \star \quad \star$$

**Exercise 5.15.** Explain why the tableau given in [U.C., (5.15)] is a special case of the general simplex tableau in [U.C., (5.22)]. Why are the last $n$ entries in the $\bar{b}$-column of [U.C., (5.15)] the coordinates of the current vertex?

**Solution 5.15.** Let $A$ be an $m \times n$ matrix and $b \geq 0$ a vector in $\mathbb{R}^m$. Define

$$\bar{A} = \begin{pmatrix} A \\ -I_n \end{pmatrix} \quad \text{and} \quad \bar{b} = \begin{pmatrix} b \\ 0 \end{pmatrix}.$$

By choosing the simplicial subset $J = \{m + 1, \ldots, n + m\}$, we compute $\bar{A}_J^{-1} = -I_n$ and $b_J = 0$. This gives $x_0 = 0$ which is a vertex of $\{x \in \mathbb{R}^n \mid$

$\bar{A}x \leq \bar{b}\}$. In this specific setup, the tableau in [U.C., (5.15)] and [U.C., (5.22)] are equivalent.

Given a simplicial subset $J$ associated to a vertex $x_0$, it follows from [U.C., (5.22)] that the lower right part of the simplex tableau is

$$\bar{b} - \bar{A}\bar{A}_J^{-1}\bar{b}_J = \begin{pmatrix} b \\ 0 \end{pmatrix} - \begin{pmatrix} -A \\ I_n \end{pmatrix} \bar{A}_J^{-1}\bar{b}_J = \begin{pmatrix} b + A\bar{A}_J^{-1}\bar{b}_J \\ -\bar{A}_J^{-1}\bar{b}_J \end{pmatrix} \tag{5.10}$$

in the notation of [U.C., (5.15)]. Due to the fact that $Q = \{x \in \mathbb{R}^n \mid (-\bar{A})x \leq \bar{b}\}$ we have $x_0 = -\bar{A}_J^{-1}\bar{b}_J$, which is the last $n$ entries in (5.10).

⋆   ⋆   ⋆

**Exercise 5.16.** Give a proof of Proposition 5.23.

**Solution 5.16.** Without loss of generality, assume $j = 1$. By assumption we have the relation

$$\tilde{A} = \begin{pmatrix} \lambda_1 & \lambda_2 & \cdots & \lambda_n \\ 0 & 1 & \cdots & 0 \\ \vdots & \vdots & \ddots & \vdots \\ 0 & 0 & \cdots & 1 \end{pmatrix} A,$$

and since $\lambda \neq 0$ and $A$ is invertible this implies

$$\tilde{A}^{-1} = A^{-1} \begin{pmatrix} \lambda_1 & \lambda_2 & \cdots & \lambda_n \\ 0 & 1 & \cdots & 0 \\ \vdots & \vdots & \ddots & \vdots \\ 0 & 0 & \cdots & 1 \end{pmatrix}^{-1} = A^{-1} \begin{pmatrix} \lambda_1^{-1} & -\lambda_2\lambda_1^{-1} & \cdots & -\lambda_n\lambda_1^{-1} \\ 0 & 1 & \cdots & 0 \\ \vdots & \vdots & \ddots & \vdots \\ 0 & 0 & \cdots & 1 \end{pmatrix}.$$

Multiplying $A^{-1}$ by this matrix corresponds to the operations mentioned in Proposition 5.23.

⋆   ⋆   ⋆

**Exercise 5.17.** Compute an optimal solution of the linear program

$$\max \quad x_1 + x_2 + x_3$$

subject to

$$
\begin{aligned}
-x_1 + x_2 + x_3 &\leq 1 \\
x_1 - x_2 + x_3 &\leq 2 \\
x_1 + x_2 - x_3 &\leq 3 \\
x_1 \quad\quad &\geq 0 \\
x_2 \quad &\geq 0 \\
x_3 &\geq 0
\end{aligned}
$$

using the simplex algorithm. Is your optimal solution unique?

**Solution 5.17.** The tableau associated with the vertex $(0,0,0)$ is

| 1 | 1 | 1 | 0 |
|---|---|---|---|
| 1 | −1 | −1 | 1 |
| (−1) | 1 | −1 | 2 |
| −1 | −1 | 1 | 3 |
| 1 | 0 | 0 | 0 |
| 0 | 1 | 0 | 0 |
| 0 | 0 | 1 | 0 |

Since $\bar{c}_1 > 0$, we choose the first column as the pivot column, and the second row becomes our pivot row cf. [U.C., (5.16)]. After normalization we apply elementary column operations to obtain the new tableau

| −1 | 2 | 0 | 2 |
|---|---|---|---|
| −1 | 0 | −2 | 3 |
| 1 | 0 | 0 | 0 |
| 1 | −2 | 2 | 1 |
| −1 | 1 | −1 | 2 |
| 0 | 1 | 0 | 0 |
| 0 | 0 | 1 | 0 |

Continuing in this way until $\bar{c}_i \leq 0$ for $i = 1, 2, 3$, we end up with the final tableau

| −1 | −1 | −1 | 6 |
|---|---|---|---|
| 0 | 0 | 1 | 0 |
| 1 | 0 | 0 | 0 |
| 0 | 1 | 0 | 0 |
| $-\frac{1}{2}$ | $-\frac{1}{2}$ | 0 | $\frac{5}{2}$ |
| 0 | $-\frac{1}{2}$ | $-\frac{1}{2}$ | 2 |
| $-\frac{1}{2}$ | 0 | $-\frac{1}{2}$ | $\frac{3}{2}$ |

According to the discussion in Exercise 5.15 we can read off the optimal vertex as the last three coordinates in the $\bar{b}$ part, meaning that the maximum is attained at the vertex $x_0 = (5/2,\ 2,\ 3/2)$. We will now argue that $x_0$ is the unique optimal point, which is a consequence of the fact that the upper left part of the simplex tableau only has negative entrances.

The linear program we have considered is $\max\{c^t x \mid x \in P\}$ where $P$ is the associated polyhedron and $c = (1,1,1)$. The upper left entrances in the last tableau show that for any $i$ we have $c^t r_i = -1 < 0$, where $r_i$ is the $i$-th extreme ray in the simplicial cone $C_J$ that contains the corresponding vertex cone $C_{x_0}$ (using notation and terminology as in §5.4.3-4). Here $J = \{1,2,3\}$. From this we use the inclusion in Lemma 5.18 to conclude that if $y \in P$, then $y = x_0 + \alpha_1 r_1 + \alpha_2 r_2 + \alpha_3 r_3$ for some $\alpha_1, \alpha_2, \alpha_3 \geq 0$. In particular, this shows that $c^t y = c^t x_0$ implies $x_0 = y$.

$$\star \quad \star \quad \star$$

**Exercise 5.18.** Let $P(\lambda)$ denote the linear program

$$\max \quad x_1 + \lambda x_2$$

subject to

$$\tfrac{1}{2}x_1 + x_2 \leq 5$$
$$x_1 + x_2 \leq 6$$
$$2x_1 - x_2 \leq 4$$
$$x_1 \qquad\quad \geq 0$$
$$x_2 \geq 0$$

for $\lambda \in \mathbb{R}$. Compute an optimal solution for $P(0)$. Find the optimal solutions for $P(\lambda)$ in general for $\lambda \in \mathbb{R}$. The problem in this exercise falls under the heading *parametric linear programming*. This means solving linear programs depending on a parameter $\lambda$.

**Solution 5.18.** In the following we will only solve the general linear program $P(\lambda)$ for $\lambda \in \mathbb{R}$, which in particular gives the solution to $P(0)$. We start by considering the vertex $(0,0)$ resulting in the tableau

| 1 | $\lambda$ | 0 |
|---|---|---|
| $-\frac{1}{2}$ | $-1$ | 5 |
| $-1$ | $-1$ | 6 |
| $\boxed{-2}$ | 1 | 4 |
| 1 | 0 | 0 |
| 0 | 1 | 0 |

$\longrightarrow$

| $-\frac{1}{2}$ | $\lambda + \frac{1}{2}$ | 2 |
|---|---|---|
| $\frac{1}{4}$ | $-\frac{5}{4}$ | 4 |
| $\frac{1}{2}$ | $-\frac{3}{2}$ | 4 |
| 1 | 0 | 0 |
| $-\frac{1}{2}$ | $\frac{1}{2}$ | 2 |
| 0 | 1 | 0 |

For $\lambda \leq -1/2$, the tableau after the first iteration shows that the vertex $(2,0)$ is optimal. Otherwise if $\lambda \geq -1/2$, we pivot around the second column and third row and find

| $\frac{\lambda-1}{3}$ | $-\frac{2\lambda+1}{3}$ | $\frac{8\lambda+10}{3}$ |
|---|---|---|
| $-\frac{1}{6}$ | $\frac{5}{6}$ | $\frac{2}{3}$ |
| 0 | 1 | 0 |
| 1 | 0 | 0 |
| $-\frac{1}{3}$ | $-\frac{1}{3}$ | $\frac{10}{3}$ |
| $\frac{1}{3}$ | $-\frac{2}{3}$ | $\frac{8}{3}$ |

From this tableau we read the vertex $(10/3, 8/3)$ to be optimal for $-1/2 \leq \lambda \leq 1$. If $\lambda \geq 1$, we continue the simplex algorithm and get the new tableau

| $2(1-\lambda)$ | $\lambda - 2$ | $4\lambda + 2$ |
|---|---|---|
| 1 | 0 | 0 |
| 0 | 1 | 0 |
| $-6$ | 5 | 4 |
| 2 | $-2$ | 2 |
| $-2$ | 1 | 4 |

This produces an optimal vertex $(2,4)$ as long as $1 \leq \lambda \leq 2$. Finally for $\lambda \geq 2$, we perform another iteration of the simplex method to find

| $-\lambda$ | $\frac{2-\lambda}{2}$ | $5\lambda$ |
|:---:|:---:|:---:|
| 1 | 0 | 0 |
| 1 | $-\frac{1}{2}$ | 1 |
| $-1$ | $-\frac{5}{2}$ | 9 |
| 0 | 1 | 0 |
| $-1$ | $-\frac{1}{2}$ | 5 |

and read $(0, 5)$ as optimal.

Note that all the solutions found are unique when $\lambda$ is in the interior of the intervals (see end of Solution 5.17), but when $\lambda$ is an endpoint there are several solutions. For example when $\lambda = 1$, every point in $\mathrm{conv}(\{(10/3,\ 8/3), (2,\ 4)\})$ is optimal.

# Chapter 6

# Closed convex subsets and separating hyperplanes

## 6.1 Introduction

In the following exercises the interest will to a great extent be on closed (and open) convex sets and the practice of finding supporting hyperplanes for such sets. For instance, supporting hyperplanes for a set in $\mathbb{R}^2$ are easily found by drawing but when considering $\mathbb{R}^d$, $d > 2$, we will face a more subtle task. It is definitely a matter of preferences how to find hyperplanes, but it may be beneficial to see how to proceed analytically and for this reason, we take this approach. Throughout the solutions, terms as open and closed sets, convergence of sequences, etc. will be used repeatedly, so if the reader feels uncomfortable with these it is advisable to consult Appendix A before continuing.

**Theorem 6.2.** *Let $C \subseteq \mathbb{R}^d$ be a closed convex subset and $y \in \mathbb{R}^d$. Then there exists a point $x_0 \in C$ closest to $y$ i.e.,*

$$|x_0 - y| \leq |x - y|$$

*for every $x \in C$. This point is uniquely given by the property that*

$$(x - x_0)^t(x_0 - y) \geq 0 \tag{6.1}$$

*for every $x \in C$.*

For a cone $C \subseteq \mathbb{R}^d$, the polar given $C^\circ$ is given as

$$C^\circ = \{\alpha \in \mathbb{R}^d \mid \alpha^t x \leq 0, \text{ for every } x \in C\}.$$

**Lemma 6.7.** *Suppose that $C \subseteq \mathbb{R}^d$ is a closed convex cone. Then*

*(1) If $x \notin C$, there exists $\alpha \in C^\circ$ with $\alpha^t x > 0$.*

*(2)* $(C^\circ)^\circ = C$.

We recall that for an affine hyperplane $H = \{x \in \mathbb{R}^n \mid \alpha^t x = \beta\}$, $H^+ = \{x \in \mathbb{R}^n \mid \alpha^t x \geq \beta\}$.

**Definition 6.10.** A *supporting hyperplane* for a convex subset $C \subseteq \mathbb{R}^d$ at a boundary point $z \in \partial C$ is an affine hyperplane $H$ with $z \in H$ and $C \subseteq H^+$.

**Theorem 6.11.** *Let $C \subseteq \mathbb{R}^d$ be a convex subset and $z \in \partial C$. Then there exists a supporting hyperplane for $C$ at $z$.*

**Theorem 6.12 (Minkowski).** *Let $K$ be a compact convex subset of $\mathbb{R}^d$. Then $\mathrm{ext}(K) \neq \emptyset$ and*

$$K = \mathrm{conv}(\mathrm{ext}(K)).$$

We recall that two sets $S_1$ and $S_2$ of $\mathbb{R}^d$ are separated by an affine hyperplane $H$ if $S_1 \subseteq H^-$ and $S_2 \subseteq H^+$. The separation is said to be proper if $S_1 \cup S_2 \not\subseteq H$, strict if $S_1 \cap H = \emptyset$ and $S_2 \cap H = \emptyset$, and strong if there exists $\epsilon > 0$, such that $S_1 + \epsilon B$ and $S_2 + \epsilon B$ are strongly separated (where $B$ denotes the unit ball in $\mathbb{R}^d$).

**Theorem 6.15 (Minkowski).** *Let $C_1, C_2 \subseteq \mathbb{R}^d$ be disjoint convex subsets. Then there exists a separating hyperplane*

$$H = \{x \in \mathbb{R}^d \mid \alpha^t x = \beta\}$$

*for $C_1$ and $C_2$. If in addition, $C_1$ is compact and $C_2$ is closed, then $C_1$ and $C_2$ can be strongly separated.*

## 6.2   Exercises and solutions

**Exercise 6.1.** Prove that

$$K = \{(x, y) \in \mathbb{R}^2 \mid x > 0, \, xy \geq 1\}$$

is a convex and closed subset of $\mathbb{R}^2$.

**Solution 6.1.** Let $(x_1, y_1), (x_2, y_2) \in K$ and $\lambda \in (0, 1)$. Then

$$(1 - \lambda)x_1 + \lambda x_2 > 0 \tag{6.2}$$

and

$$((1 - \lambda)x_1 + \lambda x_2)((1 - \lambda)y_1 + \lambda y_2)$$
$$= (1 - \lambda)^2 x_1 y_1 + \lambda^2 x_2 y_2 + (x_1 y_2 + x_2 y_1)(1 - \lambda)\lambda \qquad (6.3)$$
$$\geq (1 - \lambda)^2 + \lambda^2 + (x_1 y_2 + x_2 y_1)(1 - \lambda)\lambda$$

since $x_i y_i \geq 1$, $i = 1, 2$. Set $t = x_1/x_2 > 0$ and note that

$$x_1 y_2 + x_2 y_1 \geq \frac{x_1}{x_2} + \frac{x_2}{x_1} = t + \frac{1}{t} \geq 2$$

since $t^2 + 1 - 2t = (1 - t)^2 \geq 0$. Consequently, it follows from (6.3) that

$$((1 - \lambda)x_1 + \lambda x_2)((1 - \lambda)y_1 + \lambda y_2) \geq (1 - \lambda)^2 + \lambda^2 + 2(1 - \lambda)\lambda = 1,$$

which together with (6.2) verifies that $K$ is a convex subset. Now consider a sequence with $(x_n, y_n) \to (x, y)$ and $(x_n, y_n) \in K$. We must show that $x > 0$ and $xy \geq 1$. As $x_n \to x$ and $y_n \to y$, Proposition A.9 implies $x_n y_n \to xy$. Furthermore $x_n y_n \geq 1$ for every $n \in \mathbb{N}$, so we have

$$xy = \lim_{n \to \infty} x_n y_n \geq 1$$

and $x > 0$ that is, $(x, y) \in K$.

$$\star \quad \star \quad \star$$

**Exercise 6.2.** Give examples of convex subsets that are

    (i) bounded
    (ii) unbounded
    (iii) closed
    (iv) open
    (v) neither open nor closed

in $\mathbb{R}^2$.

**Solution 6.2.**     (i) Let $v \in \mathbb{R}^2$. The set $\{v\}$ is both convex and bounded.
    (ii) A basic example of an unbounded convex subset of $\mathbb{R}^2$ is $\mathbb{R}^2$ itself.
    (iii) The sets from (i) and (ii) are also closed and thus, they constitute examples. Alternatively, for any given $a_1 < b_1$ and $a_2 < b_2$, the set

$$[a_1, b_1] \times [a_2, b_2] = \{(x, y) \in \mathbb{R}^2 \mid a_1 \leq x \leq b_1,\ a_2 \leq y \leq b_2\}$$

is closed and convex.

(iv) The set from (ii) applies here and it holds for sets of the form

$$(a_1, b_1) \times (a_2, b_2) = \{(x, y) \in \mathbb{R}^2 \mid a_1 < x < b_1, \ a_2 < y < b_2\}$$

for $a_1 < b_1$ and $a_2 < b_2$ as well.

(v) By adjusting some of the sets above we obtain examples of this case. For instance,

$$(a_1, b_1] \times (a_2, b_2] = \{(x, y) \in \mathbb{R}^2 \mid a_1 < x \leq b_1, \ a_2 < y \leq b_2\}$$

for any choice of $a_1 < b_1$ and $a_2 < b_2$.

$$\star \quad \star \quad \star$$

**Exercise 6.3.** Let $x_0, x_0', y \in \mathbb{R}^n$. Prove that if

$$(x_0' - x_0)^t (x_0 - y) \geq 0$$
$$(x_0 - x_0')^t (x_0' - y) \geq 0,$$

then $x_0 = x_0'$.

**Solution 6.3.** The result follows from the calculation

$$\begin{aligned}
-|x_0' - x_0|^2 &= (x_0' - x_0)^t (x_0 - x_0') \\
&= (x_0' - x_0)^t (x_0 - y) + (x_0' - x_0)^t (y - x_0') \\
&= (x_0' - x_0)^t (x_0 - y) + (x_0 - x_0')^t (x_0' - y) \\
&\geq 0.
\end{aligned}$$

$$\star \quad \star \quad \star$$

**Exercise 6.4.** Can you prove that

(i) a polyhedral cone is a closed subset

(ii) a finitely generated cone is a closed subset

from first principles i.e., only using the definitions?

**Solution 6.4.** (i) Let $A$ be an $m \times d$ matrix and consider the polyhedral cone $C_1 = \{x \in \mathbb{R}^d \mid Ax \leq 0\}$. Suppose that $(x_n)_{n \in \mathbb{N}} \subseteq C_1$ is a convergent sequence with $x_n \to x$. If we let $a_i$ denote the $i$-th row of $A$, it follows by continuity of $z \mapsto a_i z$ that $a_i x \leq 0$ since $a_i x_n \leq 0$ for every $n \in \mathbb{N}$. This holds for every $i$, meaning that $x \in C_1$ and $C_1$ is closed.

(ii) A natural idea is to write a finitely generated cone as the union of cones generated by linear independent vectors, since a finite union of closed sets is closed. As long as dependence is present, one may have a hard time proving the statement. However, the content of Theorem 3.14 is exactly that it suffices to consider the case where the generators are linearly independent, thus our proof relies on this reference rather than a proof from scratch.

Consider $\mathrm{cone}(\{v_1, \ldots, v_m\})$ where $v_1, \ldots, v_m \in \mathbb{R}^d$ are linearly independent. Then we may find a $d \times d$ matrix $A$ such that

$$AV = \begin{pmatrix} I_m \\ 0 \end{pmatrix},$$

where $0$ is the $(d-m) \times m$ matrix of zeros and $V$ is the $d \times m$ matrix with columns $v_1, \ldots, v_m$. Now let $(u_n)_{n \in \mathbb{N}}$ be a convergent sequence with $u_n \to u$ and represent it as $u_n = Vr_n$ for some $r_n \in \mathbb{R}^m$ satisfying $r_n \geq 0$. Then $Au_n \to Au$, from which we deduce the existence of an $r \geq 0$ such that $r_n \to r$. We conclude that $u = Vr$, showing that $\mathrm{cone}(\{v_1, \ldots, v_m\})$ is closed.

In light of Theorems 4.12 and 4.22 it is interesting how the proofs of (i) and (ii) differ in their level of complexity. In particular compared to a finitely generated cone, it is much easier to prove that an intersection of half spaces is closed.

$$\star \quad \star \quad \star$$

**Exercise 6.5.** Let $F_1, F_2 \subseteq \mathbb{R}^2$ be closed subsets. Is

$$F_1 - F_2 = \{x - y \mid x \in F_1,\, y \in F_2\}$$

a closed subset of $\mathbb{R}^2$?

**Solution 6.5.** We give an example where $F_1 - F_2$ is not closed in $\mathbb{R}$; a generalization to $\mathbb{R}^d$ can be made by letting the $d - 1$ last coordinates be zero. Consider the two sets

$$F_1 = \left\{ m + \tfrac{1}{2m} \mid m = 1, 2, \ldots \right\}$$
$$F_2 = \{1, 2, \ldots\}$$

and a convergent sequence $(a_n)_{n \in \mathbb{N}} \subseteq F_1$. Since any convergent sequence is bounded, $(a_n)_{n \in \mathbb{N}} \subseteq \{m + \tfrac{1}{2m} \mid m = 1, 2, \ldots, M\}$ for some $M \in \{1, 2, \ldots\}$, and it follows that $(a_n)_{n \in \mathbb{N}}$ is constantly equal to an element of $F_1$ for sufficiently large $n$, thus $F_1$ is a closed subset of $\mathbb{R}$. A similar argument shows that $F_2$ is a closed subset too. However, the sequence $(b_n)_{n \in \mathbb{N}} \subseteq F_1 - F_2$ given by

$$b_n = n + \tfrac{1}{2n} - n = \tfrac{1}{2n}$$

is convergent with $\lim_{n \to \infty} b_n = 0$ which is not an element of $F_1 - F_2$, so this set is not closed.

★   ★   ★

**Exercise 6.6.** In Exercise 2.11 it is proved that

$$f(x) = (\lambda_0, \lambda_1, \dots, \lambda_d)$$

is a well defined affine map $f : \mathbb{R}^d \to \mathbb{R}^{d+1}$, where

$$x = \lambda_0 v_0 + \cdots + \lambda_d v_d$$

with $v_0, \dots, v_d$ affinely independent points in $\mathbb{R}^d$ and $\lambda_0 + \cdots + \lambda_d = 1$. Use this along with the open set

$$U = \{(\lambda_0, \dots, \lambda_d) \in \mathbb{R}^{d+1} \mid \lambda_0, \dots, \lambda_d > 0\}$$

to prove that

$$\frac{1}{d+1}(v_0 + v_1 + \cdots + v_d)$$

is an interior point of $S$ in $\mathbb{R}^d$, where $S$ is the $(d+1)$-simplex

$$S = \operatorname{conv}(\{v_0, v_1, \dots, v_d\}).$$

**Solution 6.6.** First notice that

$$z = \frac{1}{d+1}(v_0 + v_1 + \cdots + v_d) \in f^{-1}(U) = \{x \in \mathbb{R}^d \mid f(x) \in U\}$$

and that

$$\{(\lambda_0, \lambda_1, \dots, \lambda_d) \in \mathbb{R}^{d+1} \mid \lambda_0, \lambda_1, \dots, \lambda_d \geq 0\} = \overline{U}$$

shows $S = f^{-1}(\overline{U}) \supseteq f^{-1}(U)$. Next, Lemma A.21 implies that $f^{-1}(U)$ is open, and therefore $z$ is an interior point of $S$.

★   ★   ★

**Exercise 6.7.** Prove that $\operatorname{conv}(\{v_1, \ldots, v_m\})$ is a compact subset of $\mathbb{R}^n$ by considering the continuous map $f : \mathbb{R}^m \to \mathbb{R}^n$ given by

$$f(t_1, \ldots, t_m) = t_1 v_1 + \cdots + t_m v_m$$

along with $f(\Delta)$, where $\Delta = \{(t_1, \ldots, t_m)^t \in \mathbb{R}^m \mid t_1 + \cdots + t_m = 1, t_1 \geq 0, \ldots, t_m \geq 0\}$.

Hint: seek inspiration in §A.7.

**Solution 6.7.** According to Theorem A.22 in §A.7 it suffices to show that $\Delta$ is a compact subset of $\mathbb{R}^m$ since $f(\Delta) = \operatorname{conv}(\{v_1, \ldots, v_m\})$. It is not too difficult to realize that $\Delta$ is closed since it is a polyhedron (see the proof of Exercise 6.4(i)). In addition, $\Delta$ is bounded since any vector $t \in \Delta$ satisfies that $|t| \leq \sqrt{m}$.

$\star \quad \star \quad \star$

**Exercise 6.8.** Let $K$ be a compact subset of $\mathbb{R}^n$. Prove that $\operatorname{conv}(K)$ is compact by using Corollary 3.15. You may find it useful to look at Exercise 6.7.

**Solution 6.8.** Recall that

$$\Delta = \{(t_1, \ldots, t_{n+1}) \in \mathbb{R}^{n+1} \mid t_1 + \cdots + t_{n+1} = 1, t_1, \ldots, t_{n+1} \geq 0\}.$$

Consider $f : K^{n+1} \times \mathbb{R}^{n+1} \to \mathbb{R}^n$ given by

$$f(v, \lambda) = \lambda_1 v_1 + \cdots + \lambda_{n+1} v_{n+1},$$

where $v = (v_1, \ldots, v_{n+1}) \in K^{n+1}$ and $\lambda = (\lambda_1, \ldots, \lambda_{n+1}) \in \mathbb{R}^{n+1}$. (Here we use the notation $K^{n+1} = \{(k_1, \ldots, k_{n+1}) \mid k_1, \ldots, k_{n+1} \in K\}$.) From Corollary 3.15 we know that any element of $\operatorname{conv}(K)$ can be written as a convex linear combination of $n + 1$ elements and thus, we deduce that

$$f(K^{n+1} \times \Delta) = \operatorname{conv}(K).$$

Since the Cartesian product of compact sets remains compact we have that $K^{n+1} \times \Delta$ is compact, and now the result follows from Theorem A.22 by continuity of $f$.

$\star \quad \star \quad \star$

**Exercise 6.9.** Let $C \subseteq \mathbb{R}^d$ be a convex subset. Prove that $\overline{C} \neq \mathbb{R}^d$ if $C \neq \mathbb{R}^d$.

**Solution 6.9.** Suppose that $C \neq \mathbb{R}^d$. The idea is to construct a subset $S \subseteq \mathbb{R}^d$ which has the property that $C \cap S = \emptyset$ and with an interior point $x$. Then, since there exists $\epsilon > 0$ such that $|x - y| > \epsilon$ for every $y \in C$, we can conclude that $x \notin \overline{C}$. It should be noted that the structure of this argument is strongly inspired by the proof of Lemma 6.9, and the reader is encouraged to look at the nice drawing in this lemma to get a grasp of the idea.

If $\dim C < d$, $C$ is contained in an affine hyperplane $H$. Since $H$ is closed we have that $\overline{C} \subseteq H$ and the result follows. If $\dim C = d$, choose $d + 1$ affinely independent vectors $v_0, \ldots, v_d \in \mathbb{R}^d$ and let $z \notin C$. Fix $\delta > 0$ and define the $d$-simplex

$$S = \mathrm{conv}(\{(1 + \delta)z - \delta v_0, \ldots, (1 + \delta)z - \delta v_d\}).$$

If $v \in S \cap C$, there exists $\lambda_0, \ldots, \lambda_d \geq 0$ with $\lambda_0 + \cdots + \lambda_d = 1$ such that

$$v = \lambda_0((1 + \delta)z - \delta v_0) + \cdots + \lambda_d((1 + \delta)z - \delta v_d).$$

This would imply

$$z = \frac{\delta}{1 + \delta}(\lambda_0 v_0 + \cdots + \lambda_d v_d) + \frac{1}{1 + \delta}v$$

and therefore $z \in C$ by the convexity of $C$. This is a contradiction and we conclude that $S \cap C = \emptyset$. Finally, Exercise 6.6 gives that $S$ has an interior point which ends the proof.

⋆    ⋆    ⋆

**Exercise 6.10.** Find all the supporting hyperplanes of the triangle with vertices $(0,0), (0,2)$ and $(1,0)$.

**Solution 6.10.** Let $x_1 = (0, 0)$, $x_2 = (0, 2)$, and $x_3 = (1, 0)$ and define $C = \mathrm{conv}(\{x_1, x_2, x_3\})$. If we let $H_s(x_i)$ denote the set of $(\alpha_1, \alpha_2, \beta) \in \mathbb{R}^3$ for which $\{x \in \mathbb{R}^2 \mid (\alpha_1, \alpha_2)x = \beta\}$ is a supporting hyperplane for $C$ at $x_i$, $i = 1, 2, 3$, we compute that

$$H_s(x_1) = \left\{ (\alpha_1, \alpha_2, \beta) \in \mathbb{R}^3 \,\middle|\, \begin{array}{l} (\alpha_1, \alpha_2)x_1 = \beta \\ (\alpha_1, \alpha_2)x_2 \geq \beta \\ (\alpha_1, \alpha_2)x_3 \geq \beta \end{array} \right\}$$

$$= \{(\alpha_1, \alpha_2, \beta) \in \mathbb{R}^3 \mid \alpha_1, \alpha_2 \geq 0, \ \beta = 0\}$$

$$H_s(x_2) = \{(\alpha_1, \alpha_2, \beta) \in \mathbb{R}^3 \mid \alpha_1 - 2\alpha_2 \geq 0, \ \beta = 2\alpha_2 \leq 0\}$$

$$H_s(x_3) = \{(\alpha_1, \alpha_2, \beta) \in \mathbb{R}^3 \mid \alpha_1 - 2\alpha_2 \leq 0, \ \beta = \alpha_1 \leq 0\}.$$

In relation to the computations above note that $\alpha^t x \geq \beta$ for every $x \in C$ is equivalent to requiring $\min\{\alpha^t x_i \mid i = 1, 2, 3\} \geq \beta$. If we choose another boundary point $z \in \partial C$, a supporting hyperplane for $C$ at $z$ is of the form $\{x \in \mathbb{R}^2 \mid \alpha^t x = \beta\}$ where $\alpha^t z = \beta$ and $\alpha^t x \geq \beta$ for every $x \in C$. For such $(\alpha, \beta) \in \mathbb{R}^3$ we find by Corollary 4.9 that $\alpha^t x_i = \beta$ for a suitable $i \in \{1, 2, 3\}$ and thus, $(\alpha, \beta) \in H_s(x_i)$ meaning that we have found all supporting hyperplanes for $C$.

$$\star \quad \star \quad \star$$

**Exercise 6.11.** Complete the proof of Corollary 6.13.

**Solution 6.11.** Set $M = c^t z$. Inspired by the procedure in Corollary 4.9 we define

$$Q = \{x \in \mathbb{R}^d \mid c^t x = M\} \cap K.$$

Theorem 6.12 gives that $\text{ext}(Q) \neq \emptyset$ since $Q$ is convex and compact. Let $x_0 \in \text{ext}(Q)$ and assume $x_0 \notin \text{ext}(K)$. Then there exists $x, y \in K$ with $x \neq y$ such that

$$x_0 = (1 - \lambda)x + \lambda y$$

for some $0 < \lambda < 1$. From this we infer the existence of an $\epsilon > 0$, such that

$$x_0 \pm \epsilon(y - x) = (1 - (\lambda \pm \epsilon))x + (\lambda \pm \epsilon)y \in K$$

since $K$ is convex. The fact that $x_0 \in Q$ ensures $c^t x_0 = M$ and by definition,

$$c^t x_0 \geq c^t(x_0 \pm \epsilon(y - x)).$$

This is satisfied if and only if $c^t(y - x) = 0$ and thus, $x_0 \pm \epsilon(y - x) \in Q$. As a consequence, the identity

$$x_0 = \tfrac{1}{2}(x_0 - \epsilon(y - x)) + \tfrac{1}{2}(x_0 + \epsilon(y - x))$$

contradicts that $x_0$ is an extreme point of $Q$. Thus, we may conclude that $x_0$ is an extreme point of $K$ and $c^t x_0 = M$.

$$\star \quad \star \quad \star$$

**Exercise 6.12.** Give an example of a non-proper separation of convex subsets.

**Solution 6.12.** Consider the subsets $A = \{(1,0)\}$ and $B = \{(-1,0)\}$. The hyperplane $H = \{(x_1, x_2) \in \mathbb{R}^2 \mid x_2 = 0\}$ separates $A$ and $B$ but $A \cup B \subseteq H$ and thus, the separation is non-proper.

⋆   ⋆   ⋆

**Exercise 6.13.** Prove that the two subsets in Figure 6.5(b) are strictly separated closed convex subsets that cannot be separated strongly.

**Solution 6.13.** We redefine the set $S_2 = \{(x,y) \in \mathbb{R}^2 \mid x > 0, \; y \leq -1/x\}$ which is a closed convex subset of $\mathbb{R}^2$. This set is the reflection of $S_1$ in the $x$-axis.

We see that $H = \{(x,y) \in \mathbb{R}^2 \mid y = 0\}$ separates $S_1$ and $S_2$ strictly. Observe that $S_1$ and $S_2$ cannot be strongly separated as $(n, 1/n) \in S_1$ and $(n, -1/n) \in S_2$.

⋆   ⋆   ⋆

**Exercise 6.14.** In Theorem 6.15 it is proved that two disjoint convex subsets $C_1$ and $C_2$ of $\mathbb{R}^n$ can be separated by an affine hyperplane $H$ i.e., $C_1 \subseteq H^-$ and $C_2 \subseteq H^+$. Can this result be strengthened to $C_1 \cap H = \emptyset$? If not, give a counterexample.

**Solution 6.14.** As a counterexample, consider

$$C_1 = \{(x,y) \in \mathbb{R}^2 \mid y > 0, \; x \leq y\} \cup \{(0,0)\}$$
$$C_2 = \{(x,y) \in \mathbb{R}^2 \mid y \leq 0, \; -x > y\}.$$

These two disjoint convex sets only have the separating hyperplane $H = \{(x,y) \in \mathbb{R}^2 \mid y = 0\}$, but this hyperplane shares elements with both $C_1$ and $C_2$ (see Figures 6.1 and 6.2).

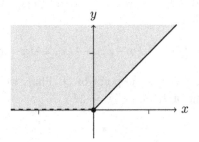

**Figure 6.1:** The set $C_1$.

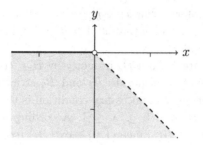

**Figure 6.2:** The set $C_2$.

$$\star \quad \star \quad \star$$

**Exercise 6.15.** Let

$$B_1 = \{(x,y) \mid x^2 + y^2 \le 1\}$$
$$B_2 = \{(x,y) \mid (x-2)^2 + y^2 \le 1\}.$$

(i) Show that $B_1$ and $B_2$ are closed convex subsets of $\mathbb{R}^2$.

(ii) Find a hyperplane properly separating $B_1$ and $B_2$.

(iii) Can you separate $B_1$ and $B_2$ strictly?

(iv) Put $B_1' = B_1 \setminus \{(1,0)\}$ and $B_2' = B_2 \setminus \{(1,0)\}$. Show that $B_1'$ and $B_2'$ are convex subsets. Can you separate $B_1'$ from $B_2$ strictly? What about $B_1'$ and $B_2'$?

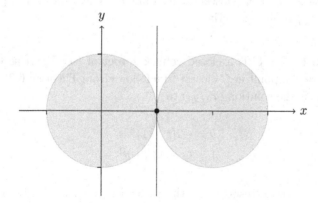

**Figure 6.3:** The sets $B_1$, $B_2$, and $H$ and the point $(1, 0)$.

**Solution 6.15.** (i) In Exercise 3.14 we showed that $B_1$ is a convex set. Now let $(x_n, y_n) \in B_1$ for $n \in \mathbb{N}$ be such that $(x_n, y_n) \to (x, y)$. By continuity it follows that $x_n^2 + y_n^2 \to x^2 + y^2$, and since $x_n^2 + y_n^2 \leq 1$ for every $n \in \mathbb{N}$ it holds that $x^2 + y^2 \leq 1$. Similarly, one shows that $B_2$ is closed and convex.

(ii) We can separate $B_1$ and $B_2$ properly by $H = \{(x, y) \in \mathbb{R}^2 \mid x = 1\}$.

(iii) It is not possible to separate $B_1$ and $B_2$ strictly since $B_1 \cap B_2 = \{(1, 0)\}$. (In general, a necessary condition is that $B_1 \cap B_2 = \emptyset$.)

(iv) Let $z_1, z_2 \in B_1'$ and $0 \leq \lambda \leq 1$. According to (i) we have $z = (1 - \lambda)z_1 + \lambda z_2 \in B_1$. Since $(1, 0) \in \text{ext}(B_1)$ (see Exercise 3.14) and $z_i \neq (1, 0)$, $i = 1, 2$, it follows that $z \neq (1, 0)$ and $z \in B_1'$. Similarly, one shows that $B_2'$ is a convex subset.

  Assume that $B_1'$ can be separated strictly from $B_2$ so that we can choose $\alpha \in \mathbb{R}^2$ and $\beta \in \mathbb{R}$ with $(x_1, y_1)\alpha < \beta$ and $(x_2, y_2)\alpha > \beta$ for every $(x_1, y_1) \in B_1'$ and $(x_2, y_2) \in B_2$. In particular, $(1, 0)\alpha > \beta$. Choose $(z_n)_{n \in \mathbb{N}} \subseteq B_1'$ with $z_n \to (1, 0)$. Then $\alpha^t z_n \to (1, 0)\alpha$ and thus, $\alpha^t z_n > \beta$ for sufficiently large $n \in \mathbb{N}$, which is a contradiction. However, $B_1'$ and $B_2'$ can be separated strictly by the affine hyperplane $H$ defined in (ii).

These conclusions are in line with Figure 6.3.

⋆ ⋆ ⋆

**Exercise 6.16.** Let $C = \{(x, y) \in \mathbb{R}^2 \mid (x - 1)^2 + y^2 \leq 1\}$ and $v = (0, 2)$. What is the point in $C$ closest to $v$? Find the equation of a hyperplane separating $\{v\}$ from $C$ strictly.

**Solution 6.16.** As $C$ is a closed convex subset of (see Exercise 6.15(i)), the existence of a point in $C$ closest to $v$ follows from Theorem 6.2. A guess of the point is the solution $(x_0, y_0)$ to

$$y = -2(x - 1)$$
$$1 = (x - 1)^2 + y^2$$

with $y_0 \geq 0$. This corresponds to the upper left point on the boundary of the disc and the line connecting $v$ and the center of the disc (see Figure 6.4). This means that our guess becomes $(x_0, y_0) = \left(1 - 1/\sqrt{5}, \, 2/\sqrt{5}\right)$. To verify that our guess was the correct one we need to show that the inequality in [U.C., (6.1)] is satisfied. Let $(x, y) \in C$. After noting that $|y| \leq \sqrt{2x - x^2}$

we compute

$$\begin{pmatrix} x - x_0 \\ y - y_0 \end{pmatrix}^t \begin{pmatrix} x_0 - 0 \\ y_0 - 2 \end{pmatrix} = \left(1 - \tfrac{1}{\sqrt{5}}\right)\left[x - \left(1 - \tfrac{1}{\sqrt{5}}\right) - 2y + \tfrac{4}{\sqrt{5}}\right]$$

$$\geq \left(1 - \tfrac{1}{\sqrt{5}}\right)\left[\sqrt{5} - 1 + \left(x - 2\sqrt{2x - x^2}\right)\right].$$

Since $(x, y) \in C$, we have that $x \in [0, 2]$, on which interval the function $z \mapsto z - 2\sqrt{2z - z^2}$ has its minimum at $1 - 1/\sqrt{5}$ with a function value of $1 - \sqrt{5}$. This verifies the optimality condition.

An affine hyperplane that separates $\{v\}$ from $C$ strictly is

$$H = \{(x, y) \in \mathbb{R}^2 \mid y = 3/2\}.$$

Note that $H$ actually separates $\{v\}$ from $C$ strongly. Such an affine hyperplane is always possible to find in our setup cf. Theorem 6.15.

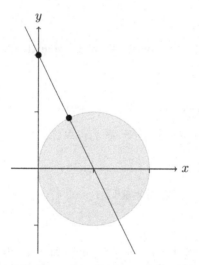

**Figure 6.4:** The set $C$, the points $(2, 0)$ and $(1 - 1/\sqrt{5}, 2/\sqrt{5})$, and the line $\{(x, y) \in \mathbb{R}^2 \mid y = -2(x - 1)\}$.

★ ★ ★

**Exercise 6.17.** Let $S$ be the square with vertices $(0,0)$, $(1,0)$, $(0,1)$ and $(1,1)$ and let $P = (2,0)$.

(i) Find the set of hyperplanes through $(1, \tfrac{1}{2})$, which separate $S$ from $P$.

(ii) Find the set of hyperplanes through $(1,0)$, which separate $S$ from $P$.

(iii) Find the set of hyperplanes through $(\frac{3}{2}, 1)$, which separate $S$ from $P$.

**Solution 6.17.** Let $H(x) \subseteq \mathbb{R}^3$ denote the set of $(\alpha_1, \alpha_2, \beta) \in \mathbb{R}^3$, $(\alpha_1, \alpha_2) \neq 0$, which we associate with a hyperplane through $x \in \mathbb{R}^3$, which separates $S$ from $P$. Furthermore, let $x_1 = (0,0)$, $x_2 = (1,0)$, $x_3 = (0,1)$, and $x_4 = (1,1)$.

(i) Necessary and sufficient conditions for checking if $(\alpha_1, \alpha_2, \beta)$ belongs to $H(1, 1/2)$ are that $(1, 1/2)\alpha = \beta$, $\alpha^t x \leq \beta$ for every $x \in S$, and $(2,0)\alpha \geq \beta$. Since $S = \text{conv}(\{x_1, x_2, x_3, x_4\})$ and $(1, 1/2) = (x_2 + x_4)/2$ the conditions translate to

$$
\begin{aligned}
\alpha^t x_2 &= \alpha_1 & &= \beta \\
\alpha^t x_4 &= \alpha_1 + \alpha_2 &&= \beta \\
\alpha^t x_1 &= & 0 &\leq \beta \\
\alpha^t x_3 &= & \alpha_2 &\leq \beta \\
(2,0)\alpha &= 2\alpha_1 && \geq \beta.
\end{aligned}
$$

Holding this together with the restriction that $\alpha \neq 0$ gives

$$H(1, 1/2) = \left\{ (\alpha_1, \alpha_2, \beta) \in \mathbb{R}^3 \mid \alpha_1 = \beta > 0, \ \alpha_2 = 0 \right\}.$$

(ii) The same procedure as in (i) shows that

$$H(1,0) = \left\{ (\alpha_1, \alpha_2, \beta) \in \mathbb{R}^3 \mid \alpha_1 = \beta \geq 0, \ \alpha_2 \leq 0, \ \alpha_1 > \alpha_2 \right\}.$$

(iii) Finally,

$$H\left(\tfrac{3}{2}, 1\right) = \left\{ (\alpha_1, \alpha_2, \beta) \in \mathbb{R}^3 \mid \beta > 0, \ \tfrac{1}{2}\beta \leq \alpha_1 \leq \beta, \ \alpha_2 = \beta - \tfrac{3}{2}\alpha_1 \right\}.$$

$\star \quad \star \quad \star$

**Exercise 6.18.** Let $K$ be a compact and convex subset of $\mathbb{R}^n$ and $C$ a closed convex cone in $\mathbb{R}^n$. Prove that

$$\text{rec}(K + C) = C.$$

**Solution 6.18.** First, recall that

$$\text{rec}(K + C) = \{d \in \mathbb{R}^n \mid x + d \in K + C \text{ for every } x \in K + C\}.$$

Since $C$ is a convex cone, $d, d' \in C$ implies $d + d' \in C$. Therefore we have that $C \subseteq \text{rec}(K + C)$. Now suppose that there exists $d \in \text{rec}(K + C) \setminus C$.

Then Lemma 6.7(1) tells us that there exists $\alpha \in C^o$ with $\alpha^t d > 0$. Since $K$ is compact, Corollary A.23 ensures the existence of $k^* \in K$ such that $\alpha^t k^* = \sup\{\alpha^t k \mid k \in K\} = \beta$. Since $\alpha \in C^o$, and by the definition of $\beta$, we find

$$\alpha^t(k + c) \leq \beta$$

for every $c \in C$ and $k \in K$. Nevertheless, $\alpha^t(k^* + d) > \beta$ implying that $k^* + d \notin K + C$. This contradicts $d \in \text{rec}(K + C)$ since $k^* \in K + C$.

$$\star \quad \star \quad \star$$

**Exercise 6.19.** A half line in $\mathbb{R}^n$ is a subset of the form $\{x + \lambda d \mid \lambda \geq 0\}$, where $d \in \mathbb{R}^n \setminus \{0\}$ and $x \in \mathbb{R}^n$. Let $C$ be a closed convex subset in $\mathbb{R}^n$. Prove that $C$ contains a half line if and only if $\text{rec}(C) \neq \{0\}$. Does this hold without the assumption that $C$ is closed?

**Solution 6.19.** Start by assuming that $\{x + \lambda d \mid \lambda \geq 0\} \subseteq C$ for $x \in C$ and $d \in \mathbb{R}^n \setminus \{0\}$. We want to show that $d \in \text{rec}(C)$. For this let $z \in C$ and note that $(x + nd)_{n \in \mathbb{N}} \subseteq C$, which implies

$$\tfrac{n-1}{n}z + \tfrac{1}{n}(x + nd) \in C$$

by convexity. Using the fact that $C$ is closed gives

$$\lim_{n \to \infty} \left(\tfrac{n-1}{n}z + \tfrac{1}{n}(x + nd)\right) = z + d \in C.$$

Assume that $\text{rec}(C) \neq \{0\}$. Exercise 3.18 shows that the recession cone is in fact a cone, thus for any $x \in C$ we have that $x + \lambda d \in C$ for some $d \neq 0$ and all $\lambda \geq 0$. This shows that $C$ contains a line.

The above result does not necessarily hold if $C$ is not closed as shown by the example given in Exercise 3.20 and the line $\{\lambda(0, 1) \mid \lambda \geq 0\}$.

$$\star \quad \star \quad \star$$

**Exercise 6.20.** Let $K \subseteq \mathbb{R}^d$ be a non-empty compact subset. Prove that $\partial K \neq \emptyset$.

**Solution 6.20.** Since $K = \overline{K}$, we need to show that

$$\overline{\mathbb{R}^d \setminus K} \setminus (\mathbb{R}^d \setminus K) = K \cap \overline{\mathbb{R}^d \setminus K} \neq \emptyset.$$

Assume for contradiction that $\mathbb{R}^d \setminus K = \overline{\mathbb{R}^d \setminus K}$. It follows by Proposition A.16 that $\mathbb{R}^d \setminus K$ is closed meaning that $K$ is open. The reader familiar with the fact that the only subsets of $\mathbb{R}^d$ that are both open and closed are $\emptyset$ and $\mathbb{R}^d$ will be done at this point. A direct argument in this case can use that $K$ is compact and $x \mapsto |x|$ is a continuous function (see Proposition A.20), so Corollary A.23 ensures that we may find $v = (v_1, \ldots, v_d) \in K$ with $|v| = \sup_{x \in K} |x| < \infty$. Without loss of generality assume that $v_1 \geq 0$. If $K$ is open as well choose $\epsilon > 0$ such that

$$v + \epsilon e_1 = (v_1 + \epsilon, v_2, \ldots, v_d) \in K.$$

However, this will lead to a contradiction by the choice of $v$ as

$$|v| = \sqrt{v_1^2 + \cdots + v_d^2} < \sqrt{(v_1 + \epsilon)^2 + v_2^2 + \cdots + v_d^2} = |v + \epsilon e_1|$$

and thus, we conclude that $\mathbb{R}^d \setminus K \neq \overline{\mathbb{R}^d \setminus K}$ and $\partial K \neq \emptyset$.

$$\star \quad \star \quad \star$$

**Exercise 6.21.** Let $C \subseteq \mathbb{R}^d$ be a convex subset. Prove that the interior $\text{int}(C) \subseteq C$ of $C$ is a convex subset.

**Solution 6.21.** Let $x, y \in \text{int}(C) \subseteq C$ and $0 < \lambda < 1$. Since $C$ is a convex subset it follows that $z = (1 - \lambda)x + \lambda y \in C \subseteq \overline{C}$. Therefore, by showing that $z \notin \partial C$ we have $z \notin \overline{\mathbb{R}^d \setminus C}$, and then $z \in \text{int}(C)$. If we suppose that $z \in \partial C$ it follows by Theorem 6.11 that there exist $\alpha = (\alpha_1, \ldots, \alpha_d) \in \mathbb{R}^d \setminus \{0\}$ and $\beta \in \mathbb{R}$ with

$$\alpha^t z = \beta \quad \text{and} \quad \alpha^t v \geq \beta \tag{6.4}$$

for every $v \in C$. In the following, we may assume that $\alpha_1 > 0$. Since $z = (1 - \lambda)x + \lambda y$ we must have that $\alpha^t x = \alpha^t y = \beta$. As $x \in \text{int}(C)$ we have $x - \epsilon e_1 \in C$ for $\epsilon > 0$ sufficiently small. (Here $e_1$ denotes the first canonical basis vector of $\mathbb{R}^d$.) At the same time,

$$\alpha^t(x - \epsilon e_1) = \alpha^t x - \epsilon \alpha_1 < \beta$$

contradicting (6.4). Therefore $z \notin \partial C$.

$$\star \quad \star \quad \star$$

**Exercise 6.22.** Show how to avoid the use of Lemma 6.9 in the proof of Theorem 6.11 if $C$ is a closed convex subset.

**Solution 6.22.** If $z \in \partial C$ then $z \in \overline{\mathbb{R}^d \setminus C}$ by definition. Consequently, there exists a sequence $(z_n)_{n \in \mathbb{N}} \subseteq \mathbb{R}^d \setminus C$ such that $z_n \to z$. Since $C = \overline{C}$, this may be used in the beginning of the proof of Theorem 6.11 instead of Lemma 6.9.

$$\star \quad \star \quad \star$$

**Exercise 6.23.** Prove Gordan's theorem by applying Theorem 6.15 to the disjoint convex subsets

$$C_1 = \{Ax \mid x \in \mathbb{R}^n\}$$

and

$$C_2 = \{y \in \mathbb{R}^m \mid y < 0\}$$

in the notation of Theorem 4.17.

**Solution 6.23.** First, notice that both conditions cannot hold at the same time since (1) implies $y^t(Ax) > 0$ while (2) implies $y^t(Ax) = (y^t A)x = 0$. Note that an equivalent formulation of (1) is that there exists $x \in \mathbb{R}^n$ such that $Ax < 0$. Assume that this is not the case such that for every $x \in \mathbb{R}^n$ there exists $i \in \{1, 2, \ldots, m\}$ with $(Ax)_i \geq 0$. Then $C_1 \cap C_2 = \emptyset$, and by Theorem 6.15 we may find $\alpha \in \mathbb{R}^m \setminus \{0\}$ and $\beta \in \mathbb{R}$ such that $H = \{x \in \mathbb{R}^m \mid \alpha^t x = \beta\}$ is a separating hyperplane for $C_1$ and $C_2$ with $C_1 \subseteq H^+$. Since $0 \in C_1 \cap \overline{C_2}$ we must have $\beta = 0$. In addition, $\alpha \geq 0$ since otherwise we may choose $y < 0$ such that $\alpha^t y > 0$. Now observe that $\alpha^t Ax \geq 0$ for every $x \in \mathbb{R}^n$ implies $\alpha^t A = 0$ (use $x = \pm e_1, \ldots, \pm e_n$), and this shows that (2) is satisfied with $y = \alpha$ in Theorem 4.17.

$$\star \quad \star \quad \star$$

**Exercise 6.24.** A compact convex subset $C \subseteq \mathbb{R}^d$ with non-empty interior $\text{int}(C) \neq \emptyset$ is called a lattice free body if $\text{int}(C) \cap \mathbb{Z}^d = \emptyset$. Give a few examples of polyhedral and non-polyhedral lattice free bodies. Prove that a lattice free body not contained in a larger lattice free body (a so-called *maximal lattice free body*) is a polyhedron.

**Solution 6.24.** A basic example of a polyhedral lattice free body in $\mathbb{R}^2$ would be the square $[0, 1]^2 = [0, 1] \times [0, 1]$ or more generally, the cube

$$[0, 1]^d = [0, 1] \times \cdots \times [0, 1]$$

in $\mathbb{R}^d$. Among non-polyhedral sets with lattice free bodies in $\mathbb{R}^2$ are discs not intersecting with $\mathbb{Z}^2$, it could be $\{(x_1, x_2) \in \mathbb{R}^2 \mid \left(x_1 - \frac{1}{2}\right)^2 + \left(x_2 - \frac{1}{2}\right)^2 \leq \frac{1}{4}\}$ or more generally, a ball

$$\left\{(x_1, \ldots, x_d) \in \mathbb{R}^d \ \Big| \ \sum_{i=1}^{d} \left(x_i - \tfrac{1}{2}\right)^2 \leq \tfrac{1}{4}\right\}$$

in $\mathbb{R}^d$. However, none of these examples are maximal lattice free bodies. An example of a set with this property in $\mathbb{R}^2$ is $\mathrm{conv}\left(\{(0, 0), (2, 0), (0, 2)\}\right)$.

The following argument shows every lattice free body is contained in a polyhedral lattice free body. We owe this clever argument to Kent Andersen. Suppose that $C \subseteq \mathbb{R}^d$ is a lattice free body. The fact that $C$ is bounded implies the existence of $M > 0$ such that $C \subseteq B_0 = [-M, M]^d$. Note that $B_0$ is a polytope. If $\mathrm{int}(B_0) \cap \mathbb{Z}^d = \emptyset$ we are done, since $C \subseteq B_0$. If $\mathrm{int}(B_0) \cap \mathbb{Z}^d \neq \emptyset$ we use the following algorithm (beginning with $i = 0$):

(1) Choose $x \in \mathrm{int}(B_i) \cap \mathbb{Z}^d$ and find a separating hyperplane $H$ such that $C \subseteq H^-$ and $x \in H^+$ (apply Theorem 6.15 to the convex subset $\mathrm{int}(C)$).

(2) Define the new polytope $B_{i+1} = B_i \cap H^-$. Notice that $C \subseteq B_{i+1}$ by construction and $x \notin \mathrm{int}(B_{i+1})$.

(3) If $\mathrm{int}(B_{i+1}) \cap \mathbb{Z}^d = \emptyset$, the algorithm terminates. If $\mathrm{int}(B_{i+1}) \cap \mathbb{Z}^d \neq \emptyset$, go to (1) with $B_{i+1}$ as a starting point.

The algorithm will terminate since the set $B_0 \cap \mathbb{Z}^d$ has to be finite as $B_0$ is bounded and $|B_{i+1} \cap \mathbb{Z}^d| < |B_i \cap \mathbb{Z}^d|$. Consequently, we end up with a polytope $B_n$ such that $\mathrm{int}(B_n) \cap \mathbb{Z}^d = \emptyset$ and $C \subseteq B_n$ for some $n \in \{0, 1, \ldots\}$.

# Chapter 7

# Convex functions

## 7.1 Introduction

Convex functions play a fundamental role in the next chapters, and it is a great advantage to be comfortable with them. Consequently, solving the following exercises is a cornerstone in the preparation for the coming material. We will see that convex functions are very well-behaved and ideally suited for minimizing over (compact) convex sets, so being able to identify if or where a function is convex is indeed useful.

Besides convexity, a considerable part of the exercises is about differentiability. For instance when a function is twice differentiable, we have a very effective criterion determining if a function is convex (see Corollary 7.20).

**Definition 7.1.** Let $C \subseteq \mathbb{R}^n$ be a convex subset. A *convex function* is a function $f : C \to \mathbb{R}$ such that

$$f((1 - \lambda)x + \lambda y) \leq (1 - \lambda)f(x) + \lambda f(y) \qquad (7.1)$$

and a *concave function* is a function $f : C \to \mathbb{R}$, such that

$$f((1 - \lambda)x + \lambda y) \geq (1 - \lambda)f(x) + \lambda f(y)$$

for every $x, y \in C$ and every $\lambda \in \mathbb{R}$ with $0 \leq \lambda \leq 1$.

A convex (concave) function $f : C \to \mathbb{R}$ is called *strictly convex* (*strictly concave*) if

$$f((1 - \lambda)x + \lambda y) = (1 - \lambda)f(x) + \lambda f(y)$$

implies that $x = y$, for every $x, y \in C$ and every $\lambda \in \mathbb{R}$ with $0 < \lambda < 1$.

99

Note that an equivalent definition of being strictly convex is that

$$f((1 - \lambda)x + \lambda y) < (1 - \lambda)f(x) + \lambda f(y)$$

for every $x, y \in C$ and $0 < \lambda < 1$ such that $x \neq y$. A similar equivalence holds for strictly concave functions.

**Lemma 7.8.** *Let $f : [a, b] \to \mathbb{R}$ be a convex function. Then*

$$\frac{f(x) - f(a)}{x - a} \leq \frac{f(b) - f(a)}{b - a} \leq \frac{f(b) - f(x)}{b - x} \qquad (7.2)$$

*for $a < x < b$.*

**Definition 7.9.** A function $f : (a, b) \to \mathbb{R}$ is *differentiable* at $x_0$ if for some $\delta > 0$ there exists a function $\epsilon : (-\delta, \delta) \to \mathbb{R}$ continuous in 0 with $\epsilon(0) = 0$ and a number $c$, such that

$$f(x_0 + h) - f(x_0) = ch + \epsilon(h)h \qquad (7.3)$$

for $x_0 + h \in (a, b)$ and $h \in (-\delta, \delta)$. The number $c$ is denoted $f'(x_0)$ and called *the derivative* of $f$ at $x_0$; $f$ is called differentiable if its derivative exists at every $x_0 \in (a, b)$.

**Definition 7.13.** A function $f : S \to \mathbb{R}$ with $S \subseteq \mathbb{R}$ is called *increasing* if

$$x \leq y \Rightarrow f(x) \leq f(y)$$

and *strictly increasing* if

$$x < y \Rightarrow f(x) < f(y)$$

for $x, y \in S$.

**Corollary 7.14.** *Let $f : (a, b) \to \mathbb{R}$ be a differentiable function. Then $f$ is increasing if and only if $f'(x) \geq 0$ for every $x \in (a, b)$. If $f'(x) > 0$ for every $x \in (a, b)$, then $f$ is strictly increasing.*

**Corollary 7.20.** *Let $f : (a, b) \to \mathbb{R}$ be a twice differentiable function. Then $f$ is convex if and only if $f''(x) \geq 0$ for every $x \in (a, b)$. If $f''(x) > 0$ for every $x \in (a, b)$, then $f$ is strictly convex.*

**Theorem 7.21.** *Let $f : (a, b) \to \mathbb{R}$ be a differentiable function. Then $f$ is convex if and only if*

$$f(y) \geq f(x) + f'(x)(y - x)$$

*for every $x, y \in (a, b)$.*

## 7.2 Exercises and solutions

**Exercise 7.1.**

(i) Prove that

$$\sqrt{ab} \leq \frac{a+b}{2}$$

for $0 \leq a \leq b$ with equality if and only if $a = b$.

(ii) Prove that $a < \sqrt{ab}$ and $(a+b)/2 < b$ for $0 < a < b$.

(iii) Start with two numbers $a$ and $b$ with $0 < a \leq b$ and define

$$a_{n+1} = \sqrt{a_n b_n}$$
$$b_{n+1} = \frac{a_n + b_n}{2},$$

where $a_0 = a$ and $b_0 = b$. Prove for $n \geq 1$ that

$$b_n - a_n < \left(\tfrac{1}{2}\right)^n (b - a)$$

if $a \neq b$.

(iv) Let $s = \lim_{n\to\infty} a_n$ and $t = \lim_{n\to\infty} b_n$. Prove, after you have convinced yourself that these limits exist, that $s = t$.

**Solution 7.1.** (i) The result follows from the inequality

$$\left(\sqrt{a} - \sqrt{b}\right)^2 \geq 0,$$

where we note that the inequality is strict if and only if $a \neq b$.

(ii) We find

$$a = \sqrt{a^2} < \sqrt{ab} \quad \text{and} \quad \frac{a+b}{2} < \frac{b+b}{2} = b.$$

(iii) First notice that (i) implies

$$0 < a_n < b_n$$

for all $n \in \mathbb{N}_0$. It follows from (ii) that

$$b_n - a_n = \frac{a_{n-1} + b_{n-1}}{2} - \sqrt{a_{n-1} b_{n-1}}$$
$$< \frac{a_{n-1} + b_{n-1}}{2} - a_{n-1} = \tfrac{1}{2} (b_{n-1} - a_{n-1})$$

for $n \geq 1$. Now one gets the result by induction.

(iv) If $a = b$ there is nothing to prove. Otherwise we combine (i) and (ii) to obtain

$$a < a_n < a_{n+1} < b_{n+1} < b_n < b$$

for $n \in \mathbb{N}$, which shows that $(a_n)_{n \in \mathbb{N}}$ is an increasing sequence bounded from above and $(b_n)_{n \in \mathbb{N}}$ is a decreasing sequence bounded from below. In light of Lemma A.11 conclude that both sequences are convergent. Now we use (iii) to write

$$0 \leq b_n - a_n < (\tfrac{1}{2})^n (b - a)$$

which shows $s = t$.

$$\star \quad \star \quad \star$$

**Exercise 7.2.** The following exercise aims at proving the inequality between the geometric and arithmetic means from scratch, without using Jensen's inequality. The proof presented here inspired Jensen's paper from 1906. It is due to Cauchy and goes back to 1821. All variables in this exercise denote non-negative real numbers.

(i) Let $a, b, c, d \geq 0$ be real numbers. Prove that

$$\sqrt[4]{abcd} \leq \frac{a + b + c + d}{4}$$

using $\sqrt[4]{x} = \sqrt{\sqrt{x}}$.

(ii) Prove by induction on $m$ that

$$\sqrt[2^m]{x_1 \cdots x_{2^m}} \leq \frac{x_1 + \cdots + x_{2^m}}{2^m}$$

for $m \geq 1$.

(iii) Suppose now that $n \in \mathbb{N}$,

$$\alpha = \frac{x_1 + \cdots + x_n}{n}$$

and $m \in \mathbb{N}$ is chosen such that $2^m \geq n$. Show that

$$\alpha = \frac{x_1 + \cdots + x_n + (2^m - n)\alpha}{2^m} \geq \sqrt[2^m]{x_1 \cdots x_n \alpha^{2^m - n}}.$$

Use this to prove the inequality between the geometric and arithmetic means.

**Solution 7.2.** (i) Using $\sqrt[4]{x} = \sqrt{\sqrt{x}}$ and $\sqrt{xy} = \sqrt{x}\sqrt{y}$ along with Exercise 7.1(i) we get

$$\sqrt[4]{abcd} \le \sqrt{\left(\frac{a+b}{2}\right)\left(\frac{c+d}{2}\right)} \le \frac{\frac{a+b}{2} + \frac{c+d}{2}}{2} = \frac{a+b+c+d}{4}.$$

(ii) The base case is covered in Exercise 7.1(i). For the inductive step we write

$$\sqrt[2^m]{x_1 \cdots x_{2^m}} = \sqrt{\sqrt[2^{m-1}]{x_1 \cdots x_{2^{m-1}}} \sqrt[2^{m-1}]{x_{2^{m-1}+1} \cdots x_{2^m}}}$$

$$\le \sqrt{\frac{x_1 + \cdots + x_{2^{m-1}}}{2^{m-1}} \cdot \frac{x_{2^{m-1}+1} + \cdots + x_{2^m}}{2^{m-1}}}$$

$$\le \frac{\frac{x_1 + \cdots + x_{2^{m-1}}}{2^{m-1}} + \frac{x_{2^{m-1}+1} + \cdots + x_{2^m}}{2^{m-1}}}{2}$$

$$= \frac{x_1 + \cdots + x_{2^m}}{2^m}.$$

(iii) Observe that

$$\alpha = \alpha + \frac{x_1 + \cdots + x_n}{2^m} - \frac{x_1 + \cdots + x_n}{2^m}$$

$$= \alpha + \frac{x_1 + \cdots + x_n}{2^m} - \frac{n\alpha}{2^m}$$

$$= \frac{x_1 + \cdots + x_n + (2^m - n)\alpha}{2^m}.$$

The numerator is a sum of $2^m$ terms, namely $x_1, \ldots, x_n$ as the first $n$ and $\alpha$ as the last $2^m - n$. Thus, we can apply what we have found in (ii) to determine that

$$\alpha = \frac{x_1 + \cdots + x_n + (2^m - n)\alpha}{2^m} \ge \sqrt[2^m]{x_1 \cdots x_n \alpha^{2^m - n}}.$$

The inequality between the geometric and arithmetic mean follows from isolating $\alpha$ in the inequality above.

$$\star \quad \star \quad \star$$

**Exercise 7.3.** Prove that

$$p(x, y) = x^2 y^2 (x^2 + y^2 - 3) + 1 \ge 0$$

for every $x, y \in \mathbb{R}$.

Hint: put $z^2 = 3 - x^2 - y^2$ for $x^2 + y^2 - 3 < 0$ and use the inequality between the geometric and arithmetic mean.

**Solution 7.3.** If $x^2 + y^2 - 3 \geq 0$ it follows that $x^2 y^2 (x^2 + y^2 - 3) + 1 \geq 0$. On the other hand if $x^2 + y^2 - 3 < 0$, define $z^2 = 3 - x^2 - y^2$. Then the inequality between the geometric and arithmetic mean yields

$$x^2 y^2 z^2 \leq \left( \frac{x^2 + y^2 + z^2}{3} \right)^2 = 1,$$

which implies $x^2 y^2 (x^2 + y^2 - 3) \geq -1$. We conclude that $p(x, y) \geq 0$.

$\star \quad \star \quad \star$

**Exercise 7.4.** Let $f : \mathbb{R} \to \mathbb{R}$ be a convex function. Prove that $f(ax + b)$ is a convex function for $a, b \in \mathbb{R}$.

**Solution 7.4.** The result is readily seen, since

$$f(a((1 - \lambda)x + \lambda y) + b) = f((1 - \lambda)(ax + b) + \lambda(ay + b))$$
$$\leq (1 - \lambda)f(ax + b) + \lambda f(ay + b).$$

$\star \quad \star \quad \star$

**Exercise 7.5.** Let $f : \mathbb{R}^n \to \mathbb{R}$ be a convex function with $f(0) = 0$. Prove that

$$-f(x) \leq f(-x)$$

for every $x \in \mathbb{R}^n$. Show that $f(x) = 0$ if $f(x) \leq 0$ for every $x \in \mathbb{R}^n$.

**Solution 7.5.** For $x \in \mathbb{R}^n$ we find by convexity of $f$ that

$$0 = f(0) = f \left( \tfrac{1}{2}x + \tfrac{1}{2}(-x) \right) \leq \tfrac{1}{2}f(x) + \tfrac{1}{2}f(-x) \tag{7.4}$$

or equivalently, $-f(x) \leq f(-x)$. If $f(x) \leq 0$, (7.4) gives $f(x) \geq -f(-x) \geq 0$ and thus, $f(x) = 0$.

$\star \quad \star \quad \star$

**Exercise 7.6.** Is $|f(x)|$ a convex function if $f(x)$ is a convex function?

**Solution 7.6.** The function $f(x) = x^2 - 1$ is convex while $|f(x)|$ is not.

$\star \quad \star \quad \star$

**Exercise 7.7.**

(i) Prove from scratch that $f(x) = x^4$ is a convex function only using that $x^2$ is a convex function.

(ii) Prove that $f(x) = x^4$ is a strictly convex function. Is $f(x) = x^6$ a strictly convex function?

(iii) Prove that $f$ is (strictly) convex if and only if $-f$ is (strictly) concave.

**Solution 7.7.**    (i) For $x, y \in \mathbb{R}$ and $0 \le \lambda \le 1$ we compute

$$((1-\lambda)x + \lambda y)^4 \le ((1-\lambda)x^2 + \lambda y^2)^2 \le (1-\lambda)x^4 + \lambda y^4,$$

where the first inequality uses that $z \mapsto z^2$ is increasing on $[0, \infty)$.

(ii) From the inequalities in (i) we see that $z \mapsto z^4$ inherits strict convexity from $z \mapsto z^2$.

By writing

$$0 \le (x^2 - y^2)^2 (x^2 + y^2) = x^6 + y^6 - x^2 y^4 - x^4 y^2$$

we conclude that $x^2 y^4 + x^4 y^2 \le x^6 + y^6$. Using this and strict convexity of $z \mapsto z^2$ and $z \mapsto z^4$ we find

$$
\begin{aligned}
((1-\lambda)x + \lambda y)^6 &< ((1-\lambda)x^4 + \lambda y^4)((1-\lambda)x^2 + \lambda y^2) \\
&= (1-\lambda)^2 x^6 + \lambda^2 y^6 + (1-\lambda)\lambda(x^2 y^4 + x^4 y^2) \\
&\le (1-\lambda)x^6 + \lambda y^6
\end{aligned}
$$

for $x \ne y$ and $0 < \lambda < 1$. This shows that $z \mapsto z^6$ is strictly convex. (Note that straightforward way of proving strict convexity is to use that $z \mapsto 6z^5$ is a strictly increasing function and apply Theorem 7.19.)

(iii) This is more or less the definition.

$$\star \quad \star \quad \star$$

**Exercise 7.8.** Let $f : [a, b] \to \mathbb{R}$ be a concave function with $f(x) > 0$ for $x \in [a, b]$. Prove that $\log f(x)$ is a concave function.

**Solution 7.8.** First, we find

$$\frac{d^2}{dx^2} \log(x) = -\frac{1}{x^2} < 0$$

for $x > 0$ and conclude by Corollary 7.20 that $z \mapsto \log(z)$ is concave on $(0, \infty)$. Let $x, y \in [a, b]$ and $0 \leq \lambda \leq 1$. Since $z \mapsto \log(z)$ is an increasing function we may write

$$\log f((1 - \lambda)x + \lambda y) \geq \log\left((1 - \lambda)f(x) + \lambda f(y)\right)$$
$$\geq (1 - \lambda)\log f(x) + \lambda \log f(y),$$

which shows that $\log \circ f : [a, b] \to \mathbb{R}$ is concave.

$$\star \quad \star \quad \star$$

**Exercise 7.9.** Let $C$ denote the set of convex functions $f : \mathbb{R} \to \mathbb{R}$.

   (i) Prove that $f + g, \lambda f \in C$ if $f, g \in C$ and $\lambda \geq 0$.
   (ii) How is the set $V$ of all functions $f : \mathbb{R} \to \mathbb{R}$ a vector space over the real numbers $\mathbb{R}$?
   (iii) Show that $C$ is a convex cone in $V$.
   (iv) Is $1 + 23412x^{10} + 23x^{48}$ a convex function?

**Solution 7.9.** Let $x, y \in \mathbb{R}$, and $\mu \in [0, 1]$.

   (i) It follows from the convexity of $f$ and $g$ that

$$f((1 - \mu)x + \mu y) + g((1 - \mu)x + \mu y)$$
$$\leq (1 - \mu)f(x) + \mu f(y) + (1 - \mu)g(x) + \mu g(y)$$
$$= (1 - \mu)(f(x) + g(x)) + \mu(f(y) + g(y)).$$

   Similarly, we find that $\lambda f((1 - \mu)x + \mu y) \leq (1 - \mu)\lambda f(x) + \mu \lambda f(y)$.
   (ii) The space $V$ is a vector space in the sense that $\lambda f \in V$ and $f + g \in V$ for every $f, g \in V$ and $\lambda \in \mathbb{R}$, and with $z \mapsto 0$ as the zero element.
   (iii) First note that $C$ is indeed a subset of $V$. The set is a cone by (i) and it is convex, since $(1 - \lambda)f + \lambda g \in C$ for $0 \leq \lambda \leq 1$.
   (iv) The function is convex by (iii), since $z \mapsto 1$, $z \mapsto z^{10}$, and $z \mapsto z^{48}$ are convex functions by Corollary 7.20.

$$\star \quad \star \quad \star$$

**Exercise 7.10.** Let $f : [a, b] \to \mathbb{R}$ be a convex function. Is $f$ a continuous function on $[a, b]$?

**Solution 7.10.** The function $f : [0, 1] \to \mathbb{R}$ given by

$$f(x) = \begin{cases} 10 & \text{if } x = 0 \\ 0 & \text{if } 0 < x \leq 1 \end{cases}$$

is convex but not continuous. (Note that a convex function will always be continuous on the open interval by Theorem 7.7.)

$$\star \quad \star \quad \star$$

**Exercise 7.11.** Sketch the graph of a differentiable function $f$, a zero $\xi$ of $f$ and a point $x_0$, such that the Newton-Raphson method fails to converge starting from $x_0$.

**Solution 7.11.** See Figure 7.1.

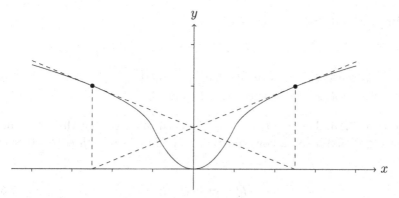

**Figure 7.1:** A graph where the Newton-Raphson method fails to converge starting from $x_0 = (5/2, 2)$ with $\xi = 0$.

$$\star \quad \star \quad \star$$

**Exercise 7.12.** Let $f : [a, b] \to \mathbb{R}$ be a twice differentiable function on $(a, b)$ with $f'' \geq 0$. Is $f : [a, b] \to \mathbb{R}$ a convex function?

**Solution 7.12.** The function

$$f(x) = \begin{cases} 1 & \text{if } x \in (a, b] \\ 0 & \text{if } x = a \end{cases}$$

is twice differentiable on $(a, b)$ and $f''(x) = 0$ but fails to be convex. (However, Corollary 7.20 gives that $f$ must be convex on the interval $(a, b)$.)

$$\star \quad \star \quad \star$$

**Exercise 7.13.** Give an example of a strictly increasing differentiable function $f : [a, b] \to \mathbb{R}$ along with $\xi \in (a, b)$ such that $f'(\xi) = 0$.

**Solution 7.13.** Consider the function $f : z \mapsto z^3$ on $[-1, 1]$. Then $f'(0) = 0$.

$$\star \quad \star \quad \star$$

**Exercise 7.14.** Give an example of a function $f : [-1, 1] \to \mathbb{R}$ differentiable on $(-1, 1)$ with $f'(0) = 0$, but where 0 is not a local extremum.

**Solution 7.14.** See Solution 7.13.

$$\star \quad \star \quad \star$$

**Exercise 7.15.** Prove in detail that the function $f(x) = |x|$ fails to be differentiable at $x = 0$. Is $f$ a convex function?

**Solution 7.15.** Let $\delta > 0$ be given. We need to argue that there does not exist $c \in \mathbb{R}$ and a continuous function $\epsilon : (-\delta, \delta) \to \mathbb{R}$ with $\epsilon(0) = 0$, such that

$$|h| = ch + \epsilon(h)h \tag{7.5}$$

for $h \in (-\delta, \delta)$. If the relation (7.5) holds, $\epsilon$ must be given by

$$\epsilon(h) = \begin{cases} 0 & \text{for } h = 0 \\ \frac{|h|}{h} - c & \text{for } h \neq 0 \end{cases}$$

$$= \begin{cases} 0 & \text{for } h = 0 \\ -(1+c) & \text{for } h \in (-\delta, 0) \\ 1 - c & \text{for } h \in (0, \delta) \end{cases} .$$

Since this function has discontinuities, we can conclude that $f$ is not differentiable at 0. The convexity of $f$ follows from Theorem A.2 (triangle inequality).

$$\star \quad \star \quad \star$$

**Exercise 7.16.** In Babylonian mathematics (3000 B.C.) one encounters the formula

$$d = h + \frac{w^2}{2h}$$

for the length of the diagonal $d$ in a rectangle with height $h$ and width $w$. Give an example showing that the formula is wrong. When is it a reasonable approximation?

(Look at the Taylor expansion for $\sqrt{1+x}$.)

**Solution 7.16.** We know from Pythagoras' theorem that

$$d_p^2 = h^2 + w^2. \tag{7.6}$$

If we let $h = w = 1$, the Babylonian formula gives $d = 3/2$ whereas $d_p = \sqrt{2}$. From (7.6) we find

$$\frac{d_p}{h} = \sqrt{1 + \frac{w^2}{h^2}}.$$

If we use a first order Taylor approximation of the function $z \mapsto \sqrt{1+z}$ around 0 we get the approximative formula

$$d_p \approx h + \frac{w^2}{2h}.$$

As a consequence, $d$ is a good approximation of $d_p$ when $\frac{w^2}{h^2}$ is small, thus the Babylonian formula is reasonable in this case.[1] As an example of this we set $h = 5$ and $w = 1$, and compute $d_p = 5.09901\ldots$ and $d = 5.1$.

$$\star \quad \star \quad \star$$

**Exercise 7.17.** Show that $f : \{x \in \mathbb{R} \mid x \geq 0\} \to \mathbb{R}$ given by

$$f(x) = -\sqrt{x}$$

is a strictly convex function.

---

[1]Note that the second derivative of $z \mapsto \sqrt{1+z}$ is bounded on any compact interval of the non-negative numbers, so the approximation accuracy follows from Theorem 7.16.

**Solution 7.17.** Since

$$f''(x) = \frac{1}{4x^{3/2}} > 0$$

for $x > 0$, Theorem 7.19 implies strict convexity of $f$ on $(0, \infty)$. The case when $x = 0$, $y > 0$, and $0 < \lambda < 1$ causes no problems since

$$f((1 - \lambda)x + \lambda y) = f(\lambda y) = -\sqrt{\lambda y} < \lambda f(y) = (1 - \lambda)f(x) + \lambda f(y).$$

$$\star \quad \star \quad \star$$

**Exercise 7.18.** Find a non-empty open subset $S \subseteq \mathbb{R}$, such that

$$f(x) = \sin(x)^2 : \mathbb{R} \to \mathbb{R}$$

is a convex function on $S$.

**Solution 7.18.** Let $S = (-\pi/4, \pi/4)$. Since $f''(x) \geq 0$ for $x \in S$, Corollary 7.20 ensures that $f$ is convex on the interval.

$$\star \quad \star \quad \star$$

**Exercise 7.19.** Let $\varphi, f : \mathbb{R} \to \mathbb{R}$ be convex functions, where $f$ is increasing. Prove that $f(\varphi(x))$ is a convex function.

**Solution 7.19.** Let $x, y \in \mathbb{R}$ and $0 \leq \lambda \leq 1$. First using that $\varphi$ is convex and $f$ is increasing, and next that $f$ convex gives

$$\begin{aligned}
f\left(\varphi((1 - \lambda)x + \lambda y)\right) &\leq f\left((1 - \lambda)\varphi(x) + \lambda\varphi(y)\right) \\
&\leq (1 - \lambda)f(\varphi(x)) + \lambda f(\varphi(y)).
\end{aligned}$$

$$\star \quad \star \quad \star$$

**Exercise 7.20.** Give conditions on $a_0, a_1, a_2, a_3, a_4 \in \mathbb{R}$ ensuring that $a_0 + a_1 x + a_2 x^2 + a_3 x^3 + a_4 x^4$ is a convex function.

**Solution 7.20.** By Corollary 7.20, a necessary and sufficient condition on $a_0, a_1, a_2, a_3, a_4 \in \mathbb{R}$ is that

$$f''(x) = 2a_2 + 6a_3 x + 12a_4 x^2 \geq 0. \tag{7.7}$$

Since this is a parabola, (7.7) is exactly satisfied when $f''$ is convex and has a non-positive discriminant that is, when $a_4 \geq 0$ and $3a_3^2 - 8a_2a_4 \leq 0$.

$$\star \quad \star \quad \star$$

**Exercise 7.21.** Let $p, q > 0$ be real numbers with $1/p + 1/q = 1$. Prove that

$$ab \leq \frac{a^p}{p} + \frac{b^q}{q}$$

for real numbers $a, b \geq 0$. This is called *Young's inequality*.

**Solution 7.21.** We may assume that $a, b > 0$. Since $x \mapsto e^x$ is a convex function we find

$$\begin{aligned} ab &= \exp\left(\tfrac{1}{p}\log(a^p) + \tfrac{1}{q}\log(b^q)\right) \\ &\leq \tfrac{1}{p}\exp\left(\log(a^p)\right) + \tfrac{1}{q}\exp\left(\log(b^q)\right) \\ &= \tfrac{1}{p}a^p + \tfrac{1}{q}b^q. \end{aligned}$$

$$\star \quad \star \quad \star$$

**Exercise 7.22.** Prove that a bounded differentiable convex function $f : \mathbb{R} \to \mathbb{R}$ is constant (bounded means that there exists $M \in \mathbb{R}$, such that $|f(x)| \leq M$ for every $x \in \mathbb{R}$).

**Solution 7.22.** Let $x \in \mathbb{R}$. Theorem 7.21 implies that for any $y \in \mathbb{R}$,

$$f(y) \geq f(x) + f'(x)(y - x).$$

If $f$ is bounded, the inequality implies $f'(x) = 0$. Since $x$ was arbitrary we conclude by Corollary 7.14 that $f$ must be constant.

$$\star \quad \star \quad \star$$

**Exercise 7.23.** Use Lemma 7.8 to do Exercise 7.22 without assuming that $f$ is differentiable.

**Solution 7.23.** Consider $x, y \in \mathbb{R}$ where $y < x$. For sufficiently large $n$, $y \in (-n, x)$ and $x \in (y, n)$, and then Lemma 7.8 gives

$$\frac{f(x) - f(-n)}{x + n} \leq \frac{f(x) - f(y)}{x - y} \leq \frac{f(n) - f(y)}{n - y}. \tag{7.8}$$

Since $f$ is bounded, the left and right side in (7.8) converges to zero as $n \to \infty$, and we conclude that $f(x) = f(y)$.

# Chapter 8

# Differentiable functions of several variables

## 8.1  Introduction

This chapter concerns several aspects of differentiability. In particular, the exercises are centered around the application of the multidimensional chain rule (Theorem 8.12) and the study of partial derivatives, where the latter is motivated by Proposition 8.4 and Theorem 8.6. In addition, considerable attention is paid to how differentiability can ease the procedure of optimizing in a suitable setup. Specifically, under the right conditions, the existence of Lagrange multipliers at an optimal point helps us in solving several practical problems.

**Theorem 8.5.** *Let $f : U \to \mathbb{R}^m$ be a function with $U \subseteq \mathbb{R}^n$ an open subset. If the partial derivatives for $f$ exist at every $x \in U$ with*

$$\frac{\partial f_i}{\partial x_j}$$

*continuous (for $i = 1, \ldots, m$ and $j = 1, \ldots, n$), then $f$ is differentiable. If the second order partial derivatives exist for a function $f : U \to \mathbb{R}$ and are continuous functions, then*

$$\frac{\partial^2 f}{\partial x_i \partial x_j} = \frac{\partial^2 f}{\partial x_j \partial x_i}$$

*for $i, j = 1, \ldots, n$.*

The next Theorem is also known as the chain rule.

**Theorem 8.12.** *Let $f : U \to \mathbb{R}^m$ and $g : V \to \mathbb{R}^n$ with $U \subseteq \mathbb{R}^n$, $V \subseteq \mathbb{R}^l$ open subsets and $g(V) \subseteq U$. If $g$ is differentiable at $x_0 \in V$ and $f$ is* *differentiable at $g(x_0) \in U$, then $f \circ g$ is differentiable at $x_0$ with*

$$(f \circ g)'(x_0) = f'(g(x_0))g'(x_0).$$

**Theorem 8.17.** *Suppose that $x_0 \in S$ is a local extremum for [U.C., (8.16)] and that*

$$\nabla g_1(x_0), \dots, \nabla g_m(x_0)$$

*are linearly independent. Then there exists $\lambda_1, \dots, \lambda_m \in \mathbb{R}$, such that $(x_0, \lambda_1, \dots, \lambda_m) \in \mathbb{R}^{n+m}$ is a critical point for the Lagrangian of [U.C., (8.16)].*

## 8.2    Exercises and solutions

**Exercise 8.1.** Consider $f : \mathbb{R} \to \mathbb{R}^3$ and $g : \mathbb{R}^3 \to \mathbb{R}$ given by

$$f(t) = \begin{pmatrix} t \\ t^2 \\ t^3 \end{pmatrix} \quad \text{and} \quad g\begin{pmatrix} x \\ y \\ z \end{pmatrix} = x^2 + 3y^6 + 2z^5.$$

Compute $(g \circ f)'(t)$ using the chain rule and check the result with an explicit computation of the derivative of $g \circ f : \mathbb{R} \to \mathbb{R}$.

**Solution 8.1.** It is seen from Theorem 8.12 that $(g \circ f)'(t) = g'(f(t))f'(t)$ for $t \in \mathbb{R}$. Since

$$f'(t) = \begin{pmatrix} 1 \\ 2t \\ 3t^2 \end{pmatrix} \quad \text{and} \quad g'(x, y, z) = (2x, 18y^5, 10z^4)$$

we find the derivative of $g \circ f$ as

$$(g \circ f)'(t) = (2t, 18(t^2)^5, 10(t^3)^4) \begin{pmatrix} 1 \\ 2t \\ 3t^2 \end{pmatrix} = 2t + 36t^{11} + 30t^{14}.$$

This agrees with the derivatives of $(g \circ f)(t) = t^2 + 3t^{12} + 2t^{15}$.

$$\star \quad \star \quad \star$$

**Exercise 8.2.** Consider the function $f : \mathbb{R} \to \mathbb{R}$ defined by

$$f(x) = \begin{cases} x^2 \sin(\frac{1}{x}), & \text{if } x \neq 0 \\ 0, & \text{if } x = 0. \end{cases}$$

Prove that $f$ is differentiable. Is $f'$ a continuous function?

**Solution 8.2.** If $x_0 \neq 0$, then the function is differentiable at $x_0$ according to Theorem 8.12. If $x_0 = 0$ we notice that for $h \in \mathbb{R}$, $f(x_0 + h) = h\epsilon(h)$, where $\epsilon$ is the continuous function given by

$$\epsilon(h) = \begin{cases} h\sin(\frac{1}{h}) & \text{if } h \neq 0 \\ 0 & \text{if } h = 0. \end{cases}$$

Hence it follows from the definition that $f$ is differentiable at $x_0 = 0$ with derivative $c = 0$. The chain rule and using $(fg)' = f'g + fg'$ (Leibniz's product rule) gives

$$f'(x) = \begin{cases} 2x\sin\left(\frac{1}{x}\right) - \cos\left(\frac{1}{x}\right) & \text{if } x \neq 0 \\ 0 & \text{if } x = 0. \end{cases}$$

This function is not continuous at $x_0 = 0$ since for every $\delta > 0$ it is possible to find $|x| \leq \delta$ such that $\sin(1/x) = 0$ and $\cos(1/x) = 1$, and thus $|f(0) - f(x)| = 1$.

$$\star \quad \star \quad \star$$

**Exercise 8.3.** Let $f : \mathbb{R}^2 \to \mathbb{R}$ be given by

$$f(x, y) = \begin{cases} \dfrac{xy(x^2 - y^2)}{x^2 + y^2} & \text{if } (x, y) \neq (0, 0) \\ 0 & \text{if } (x, y) = (0, 0). \end{cases}$$

Verify that

$$\frac{\partial f}{\partial x}(v) = \frac{y\left(x^4 + 4x^2y^2 - y^4\right)}{\left(x^2 + y^2\right)^2}$$

$$\frac{\partial f}{\partial y}(v) = \frac{x\left(x^4 - 4x^2y^2 - y^4\right)}{\left(x^2 + y^2\right)^2}$$

and show that

$$\left|\frac{\partial f}{\partial x}(v)\right| \leq 2|y|$$

$$\left|\frac{\partial f}{\partial y}(v)\right| \leq 2|x|$$

for $v = (x, y) \neq (0, 0)$. Show that

$$\frac{\partial f}{\partial x}(0, y) = \begin{cases} -y & \text{if } y \neq 0 \\ 0 & \text{if } y = 0, \end{cases}$$

$$\frac{\partial f}{\partial y}(x, 0) = \begin{cases} x & \text{if } x \neq 0 \\ 0 & \text{if } x = 0. \end{cases}$$

Prove that $f$ is a differentiable function using Theorem 8.5. Verify that

$$\frac{\partial^2 f}{\partial x \partial y}(0, 0) = 1 \quad \text{and} \quad \frac{\partial^2 f}{\partial y \partial x}(0, 0) = -1.$$

What does this have to do with the second part of Theorem 8.5?

**Solution 8.3.** The expressions for $\frac{\partial f}{\partial x}(v)$ and $\frac{\partial f}{\partial y}(v)$ can be verified using

$$\left(\frac{f}{g}\right)' = \frac{f'g - g'f}{g^2}$$

(the quotient rule). It follows from the triangle inequality (Theorem A.2) that

$$\begin{aligned} \left| \frac{\partial f}{\partial x}(v) \right| &= \left| \frac{y(x^4 + 4x^2y^2 - y^4)}{(x^2 + y^2)^2} \right| \\ &\leq \frac{|y|(x^4 + 4x^2y^2 + y^4)}{(x^2 + y^2)^2} \\ &\leq \frac{|y|(2x^4 + 4x^2y^2 + 2y^4)}{(x^2 + y^2)^2} \\ &= 2|y|. \end{aligned} \tag{8.1}$$

Similar calculations show that

$$\left| \frac{\partial f}{\partial y}(v) \right| \leq 2|x|.$$

Let $y \neq 0$. Then

$$\frac{\partial f}{\partial x}(0, y) = \frac{y(-y^4)}{(y^2)^2} = -y.$$

If $y = 0$,

$$\frac{\partial f}{\partial x}(0, 0) = \lim_{\delta \to 0} \frac{f(\delta, 0) - f(0, 0)}{\delta} = 0.$$

Similarly, $\frac{\partial f}{\partial y}(x,0) = x$ for $x \neq 0$ and $\frac{\partial f}{\partial y}(0,0) = 0$.

Observe that

$$\frac{\partial f}{\partial x}(v) = \frac{y(x^4 + 4x^2 y^2 - y^4)}{(x^2 + y^2)^2}$$

is continuous at $v \neq (0,0)$. Furthermore $\frac{\partial f}{\partial x}(v)$ is continuous at $(0,0)$ by (8.1) and consequently, $v \mapsto \frac{\partial f}{\partial x}(v)$ is a continuous function on $\mathbb{R}^2$. The continuity of $v \mapsto \frac{\partial f}{\partial y}(v)$ is a similar argument. Therefore we conclude that $f$ is differentiable by Theorem 8.5.

Now we compute the derivatives as

$$\frac{\partial}{\partial y}\left(\frac{\partial f}{\partial x}\right)(0,0) = \lim_{\delta \to 0} \frac{\frac{\partial f}{\partial x}(0,\delta) - \frac{\partial f}{\partial x}(0,0)}{\delta} = \lim_{\delta \to 0} \frac{-\delta - 0}{\delta} = -1$$

$$\frac{\partial}{\partial x}\left(\frac{\partial f}{\partial y}\right)(0,0) = \lim_{\delta \to 0} \frac{\frac{\partial f}{\partial y}(\delta,0) - \frac{\partial f}{\partial y}(0,0)}{\delta} = \lim_{\delta \to 0} \frac{\delta - 0}{\delta} = 1.$$

The second part of Theorem 8.5 then says that the second order partial derivatives cannot be continuous.

⋆ ⋆ ⋆

**Exercise 8.4.** How would you find a numerical approximation to a root of the equation $x = \cos(x)$? Here is a suggestion for a strategy you may not have seen before. First introduce an extra parameter $t$ parametrizing the family

$$x = t \cos(x)$$

of equations. Of course you know the solution for $t = 0$. You want it for $t = 1$. You can do this by following a function $x : [0,1] \to \mathbb{R}$ with $x(0) = 0$ satisfying

$$x(t) = t \cos(x(t)).$$

Show that such a function satisfies

$$x'(t) = \frac{\cos(x(t))}{1 + t\sin(x(t))}.$$

Use the approximation

$$x(t_0 + \delta) = x(t_0) + \delta x'(t_0)$$

with step size $\delta = 0.1$ in this method to find an approximate solution to $x = \cos(x)$. Do you see a way of improving this numerical method (apart from choosing a smaller $\delta$)?

**Solution 8.4.** Explicit differentiation gives

$$x'(t) = \cos(x(t)) - tx'(t)\sin(x(t))$$

and the identity follows. The first iteration yields

$$x(0 + 0.1) = x(0) + 0.1\frac{\cos(0)}{1 + 0 \cdot \sin(0)}$$

$$= 0.1.$$

Iterating twice results in

$$x(0.1 + 0.1) = x(0.1) + 0.1\frac{\cos(0.1)}{1 + 0.1 \cdot \sin(0.1)}$$

$$= 0.1985.$$

Continuing in this fashion we get

$$x(1) = 0.7606. \tag{8.2}$$

This kind of approximation will induce a bias as soon as the function $t \mapsto x(t)$ is non-linear. A way to (hopefully) reduce this bias is to note that, for a given $0 < t_0 < 1$, the above approximation of $x(t_0)$ corresponds to an approximate solution to $x - t_0 \cos(x) = 0$. Using this as a starting point in the Newton-Raphson method one may then improve the accuracy of the solution in a fixed step. This procedure can then be applied to every step of the algorithm. Alternatively, or complementary, an attempt to reduce the error is to expand further along the Taylor polynomial e.g. approximating $x(t_0 + \delta)$ by $x(t_0) + \delta x'(t_0) + \frac{1}{2}\delta^2 x''(t_0)$.

★   ★   ★

**Exercise 8.5.** Fill in the details in the computations of Example 8.19.

**Solution 8.5.** We want to find the maximum and minimum value of $f(x, y, z) = x - 2y + 2z$ when $x^2 + y^2 + z^2 = 1$ and $x + y + z = 0$. Initially, note that $f$ is optimized on a compact set so Corollary A.23 applies and $f$ will attain its minimum and maximum on this set.

With the setup in [U.C., (8.18)] we add the last three equations to get $1 = 2\lambda(x + y + z) + 3\mu$, and since $x + y + z = 0$ this implies $\mu = 1/3$. Inserting this value in the last three equations and isolating for $x$, $y$, and $z$ give

$$x = \frac{1}{3\lambda}, \quad y = -\frac{7}{6\lambda}, \quad \text{and} \quad z = \frac{5}{6\lambda}.$$

Therefore

$$\frac{1}{9\lambda^2} + \frac{49}{36\lambda^2} + \frac{25}{36\lambda^2} = 1$$

or equivalently, $\lambda = \pm\sqrt{13/6}$. The case $\lambda = \sqrt{13/6}$ is already handled in the example, and when $\lambda = -\sqrt{13/6}$ we get

$$v = \begin{pmatrix} x \\ y \\ z \end{pmatrix} = \begin{pmatrix} -\sqrt{2/39} \\ \sqrt{7/78} \\ -\sqrt{5/78} \end{pmatrix}.$$

By Theorem 8.17 the local extrema are to be found in these points and hence, they have to correspond to the (global) minimum and maximum mentioned in the beginning. The identity

$$f(-v) = -f(v) > 0$$

prevails the minimum and the maximum.

$$\star \quad \star \quad \star$$

**Exercise 8.6.** Maximize $x^2 + y^2$ subject to $x^2 + xy + y^2 = 4$. Does this problem have a geometric interpretation?

**Solution 8.6.** The equations from Remark 8.18 read

$$x^2 + xy + y^2 - 4 = 0$$
$$2x = \lambda(2x + y)$$
$$2y = \lambda(2y + x).$$

Notice $2x + y \neq 0$ and $2y + x \neq 0$, and therefore we can isolate $\lambda$ to deduce that

$$\frac{2x}{2x + y} = \frac{2y}{2y + x}.$$

This implies $x = \pm y$. Letting $x = -y$ results in the two solutions $(2, -2)$ and $(-2, 2)$ which, when compared to the case $x = y$, can be verified as the maxima.

The solution can be interpreted as the point on the ellipse $x^2 + xy + y^2 = 4$ with the largest distance to the origin.

$$\star \quad \star \quad \star$$

**Exercise 8.7.** Fill in the details in the proof of the inequality between the geometric and arithmetic means given in §8.4.

**Solution 8.7.** We start by noticing that the inequality between the geometric and arithmetic mean is always satisfied if $x_i = 0$ for some $i$, and we can therefore assume $x_i > 0$ for all $i$. If $a = 0$ then $g(x_1, \ldots, x_n) = 0$ together with $x_i \geq 0$ for $i = 1, \ldots, n$ gives that $x_1 = \cdots = x_n = 0$, and the inequality will follow in this case.

The two equations

$$x_2 x_3 \cdots x_n = \lambda \quad \text{and} \quad x_1 x_3 \cdots x_n = \lambda,$$

implies $x_1 = x_2$. Continuing this way gives $x_1 = \cdots = x_n$, and since $g(x_1, \ldots, x_n) = 0$, they must have the common value $a$.

**Exercise 8.8.** A rectangular box has side lengths $x$, $y$ and $z$. What is its maximal volume when we assume that $(x, y, z)$ lies on the plane

$$\frac{x}{a} + \frac{y}{b} + \frac{z}{c} = 1$$

for $a, b, c > 0$.

**Solution 8.8.** Due to the non-negativity of $x, y$, and $z$, we are optimizing over a compact set. By Corollary A.23, this means that the function $f : \mathbb{R}^3 \to \mathbb{R}$ where $f(x, y, z) = xyz$ must attain its maximum value. The equations from Remark 8.18 read

$$\frac{x}{a} + \frac{y}{b} + \frac{z}{c} = 1$$

$$yz = \frac{\lambda}{a}$$

$$xz = \frac{\lambda}{b}$$

$$xy = \frac{\lambda}{c}.$$

Assuming $x, y, z \neq 0$, these equations imply

$$\frac{x}{a} = \frac{y}{b} = \frac{z}{c} = \frac{1}{3}$$

which gives a volume of

$$f\left(\frac{a}{3}, \frac{b}{3}, \frac{c}{3}\right) = \frac{abc}{9}.$$

By Theorem 8.17 this is a candidate for a maximum. Any other candidate must necessarily have $x, y$, or $z$ equal to zero. However, these cannot be maximal since they have a function value of zero. By the remark in the beginning we therefore conclude that $(a/3, b/3, c/3)$ maximizes the volume of the box.

$$\star \quad \star \quad \star$$

**Exercise 8.9.** A company is planning to produce a box with volume $2\,\mathrm{m}^3$. For design reasons it needs different materials for the sides, top and bottom. The cost of the materials per square meter is 1 dollar for the sides, 1.5 dollars for the bottom and the top. Find the measurements of the box minimizing the production costs.

**Solution 8.9.** We let $w$ be the width, $h$ the height, and $d$ the depth of the box. The problem can then be formulated as a solution to

$$\min 2hw + 2hd + 3wd$$
$$\text{s.t.} \quad hwd = 2.$$

According to Remark 8.18 we need to solve the four equations

$$2h + 3d = \lambda hd$$
$$2h + 3w = \lambda hw$$
$$2w + 2d = \lambda wd$$
$$hwd = 2.$$

For example multiplying the two first equations by $w$ and $d$ respectively, and then subtracting them gives $w = d$ as $h = 0$ is not a possibility. Similar reasoning gives $h = \frac{3}{2}d$ and $d^3 = 4/3$, and we conclude that the unique solution to these equations is

$$(w, h, d, \lambda) = \left( \sqrt[3]{4/3}, \sqrt[3]{9/2}, \sqrt[3]{4/3}, 2\sqrt[3]{6} \right) \approx (1.10, 1.65, 1.10, 3.63).$$

When $hwd = 2$ we have $2hd + 2hw + 3wd \geq \frac{2}{w} + hw + \frac{2}{h}$, and since this expression explodes if the length of $(w, h, d)$ becomes large, $(w, h, d) \mapsto 2hw + 2hd + 3wd$ attains its minimum on $\{(w.h, d) \mid hwd = 2, \ h, w, d > 0\}$. Therefore the minimum must be attained in the $(w, h, d)$ found above.

$$\star \quad \star \quad \star$$

**Exercise 8.10.** What is wrong with the following proof of Theorem 8.17?

**Proof.**  Suppose there exists a non-zero vector $v \in \mathbb{R}^n$ with $\nabla f(v_0)v \neq 0$ and

$$\nabla g_1(v_0)v = \cdots = \nabla g_m(v_0)v = 0.$$

Now define one variable differentiable functions by $F(t) = f(v_0 + tv)$ and $G_i(t) = g_i(v_0 + tv)$ for $i = 1, \ldots, m$. By the chain rule one gets

$$F'(t) = \nabla f(v_0)v \neq 0$$
$$G_i'(t) = \nabla g_i(v_0)v = 0.$$

This implies that $g_i(v_0 + tv) = g_i(v_0) = 0$ for every $t \in \mathbb{R}$. Therefore $v_0$ cannot be an extremum under these conditions and it follows that $\nabla f(v_0)$ lies in the linear span of $\nabla g_1(v_0), \ldots, \nabla g_m(v_0)$.  □

**Solution 8.10.**  The problem is the application of the chain rule (Theorem 8.12). By correctly applying this rule one obtains

$$F'(t) = \nabla f(v_0 + tv)v$$
$$G_i'(t) = \nabla g_i(v_0 + tv)v,$$

which does not have the implications alluded to above.

# Chapter 9

# Convex functions of several variables

## 9.1 Introduction

The scope is on convex functions of several variable, and the exercises will be on verifying this property for specific functions, but also on closely related concepts. Indeed, since the (set of) subgradients can be used to characterize the class of convex functions (Theorem 9.3), these will be handled in a part of the exercises. In addition, one of the main result of this section (Theorem 9.5) gives, under suitable assumptions, a characterization of convexity of a function in terms of its associated Hessian matrix. This boils down to check positive (semi)definiteness of matrices and thus, there will be separate exercises for this as well. Finally, in relation to this, some exercises touch upon quadratic forms and their contour plots.

**Corollary 9.2.** *Let $f : U \to \mathbb{R}$ be a differentiable convex function, where $U$ is an open convex subset of $\mathbb{R}^n$. A point $x_0 \in U$ is a global minimum for $f$ if and only if $\nabla f(x_0) = 0$.*

**Definition 9.4.** A real symmetric matrix $A$ is called *positive definite* if

$$v^t A v > 0$$

for every $v \in \mathbb{R}^n \setminus \{0\}$ and *positive semidefinite*

$$v^t A v \geq 0$$

for every $v \in \mathbb{R}^n$.

For an open set $U$, the Hessian matrix of a twice differentiable function $f : U \to \mathbb{R}$ at $x \in U$ contains the double derivative $\frac{\partial^2 f}{\partial x_i \partial x_j}(x)$ on the $(i,j)$-th entry.

**Theorem 9.5.** *Let $f : U \to \mathbb{R}$ be a differentiable function with continuous second order partial derivatives, where $U \subseteq \mathbb{R}^n$ is a convex open subset. Then $f$ is convex if and only if the Hessian $\nabla^2 f(x)$ is positive semidefinite for every $x \in U$. If $\nabla^2 f(x)$ is positive definite for every $x \in U$, then $f$ is strictly convex.*

**Lemma 9.6.** *A diagonal matrix*

$$D = \begin{pmatrix} \lambda_1 & 0 & \cdots & 0 \\ 0 & \lambda_2 & \cdots & 0 \\ \vdots & \vdots & \ddots & \vdots \\ 0 & 0 & \cdots & \lambda_n \end{pmatrix}$$

*is positive definite if and only if $\lambda_1, \ldots, \lambda_n > 0$ and positive semidefinite if and only if $\lambda_1, \ldots, \lambda_n \geq 0$. If $B$ is an invertible matrix and $A$ is a symmetric matrix, then $A$ is positive definite (semidefinite) if and only if $B^t A B$ is positive definite (semidefinite).*

**Theorem 9.11.** *A symmetric $n \times n$ matrix $A = (a_{ij})_{1 \leq i,j \leq n}$ is positive definite if and only if its leading principal minors are $> 0$.*

**Theorem 9.15.** *A symmetric $n \times n$ matrix $A = (a_{ij})_{1 \leq i,j \leq n}$ is positive semidefinite if and only if its $2^n - 1$ principal minors are non-negative.*

**Corollary 9.18.** *Let $A$ be a real symmetric $n \times n$ matrix. Then there exists a diagonal matrix $D$ and an invertible matrix $C$, such that*

$$A = C^t D C. \tag{9.1}$$

*$A$ is positive semidefinite if and only if there exists a matrix $S$ such that $A = S^t S$. Every positive semidefinite matrix is a sum of at most $n$ positive semidefinite rank one matrices.*

**Theorem 9.28.** *Let $f : \mathbb{R}^n \to \mathbb{R}$ be a quadratic form given by*

$$f(x) = x^t C x,$$

*where $C$ is a real symmetric matrix. Then*

    *(i) $f$ is convex if and only if $C$ is positive semidefinite and $f$ is strictly convex if and only if $C$ is positive definite.*

    *(ii) There exists a basis $b_1, \ldots, b_n$ of $\mathbb{R}^n$ and numbers $\lambda_1, \ldots, \lambda_n$, such that*

$$f(x) = \lambda_1 (b_1^t x)^2 + \cdots + \lambda_n (b_n^t x)^2. \tag{9.2}$$

*(iii) The numbers $p = |\{i \mid \lambda_i > 0\}|$ and $q = |\{i \mid \lambda_i < 0\}|$ of positive and negative signs among $\{\lambda_1, \ldots, \lambda_n\}$ are independent of the chosen expression in [U.C., (9.19)].*

We end by recalling that $\xi \in \mathbb{R}^n$ is called a subgradient for a function $f : S \to \mathbb{R}$ at $x_0 \in S$ for some $S \subseteq \mathbb{R}^n$ if

$$f(x) \geq f(x_0) + \xi^t(x - x_0)$$

for every $x \in S$.

## 9.2 Exercises and solutions

**Exercise 9.1.** Show in detail that $\partial f(0) = [-1, 1]$ for $f(x) = |x|$.

**Solution 9.1.** By definition, $\xi \in \partial f(0)$ if and only if

$$|x| \geq \xi x \tag{9.3}$$

for every $x \in \mathbb{R}$. It follows that (9.3) holds for $\xi \leq 1$ if $x \geq 0$ and for $\xi \geq -1$ if $x \leq 0$. Consequently, $\partial f(0) = [-1, 1]$.

$$\star \quad \star \quad \star$$

**Exercise 9.2.** Let $S$ be a subset and $f : S \to \mathbb{R}$. Suppose that $0 \in \partial f(x_0)$. What can you say about $x_0$?

**Solution 9.2.** The statement $0 \in \partial f(x_0)$ means that $f(x) \geq f(x_0)$ for every $x \in S$ so, in other words, $x_0$ is a global minimum for $f$.

$$\star \quad \star \quad \star$$

**Exercise 9.3.** Let $U$ be a convex open subset of $\mathbb{R}^n$ and $f : U \to \mathbb{R}$ a bounded convex function (bounded means that there exists $M \in \mathbb{R}$ such that $|f(x)| \leq M$ for every $x \in U$). Prove that for every $x_0 \in U$ there exists $\epsilon > 0$ such that

$$S(\epsilon) = \{\xi \in \partial f(x) \mid |x - x_0| < \epsilon\} \subseteq \mathbb{R}^n$$

is a bounded subset.

**Solution 9.3.** Let $M \in \mathbb{R}$ be given such that $|f(x)| \leq M$ for every $x \in U$. For $x_0 \in U$ we may find $\delta > 0$, such that $|x - x_0| < \delta$ implies $x \in U$. Put $\epsilon = \delta/2$. If $|x - x_0| < \epsilon$ then

$$\left|x \pm \tfrac{\delta}{2}e_i - x_0\right| = \left|x \pm \tfrac{\delta}{2}e_i - x + x - x_0\right|$$
$$\leq \left|x \pm \tfrac{\delta}{2}e_i - x\right| + |x - x_0| < \tfrac{\delta}{2} + \epsilon = \delta$$

for a canonical basis vector $e_i$ of $\mathbb{R}^n$. Suppose now that $\xi \in \partial f(x)$. Then

$$f(y) - f(x) \geq \xi^t(y - x)$$

for every $y \in U$. For $y = x \pm \tfrac{\delta}{2}e_i$ it follows that

$$2M \geq \pm\frac{\delta}{2}\xi^t e_i.$$

This proves that $\pm\xi^t e_i \leq 4M/\delta$ for $i = 1, \ldots, n$, and therefore that $S(\epsilon)$ is a bounded subset.

$$\star \quad \star \quad \star$$

**Exercise 9.4.** This exercise describes the computation of $f_d''(t)$ using the chain rule noticing that $f_d'(t)$ is expressed in [U.C., (9.6)] as the composition of three differentiable maps:

$$g_1(t) = x_0 + td$$
$$g_2(u) = \nabla f(u)$$
$$g_3(u) = u^t d,$$

where $g_1 : \mathbb{R} \to \mathbb{R}^n$, $g_2 : \mathbb{R}^n \to \mathbb{R}^n$ and $g_3 : \mathbb{R}^n \to \mathbb{R}$. Show that $f_d'(t) = g_3 \circ g_2 \circ g_1$, where $f_d'(t)$ is given in [U.C., (9.6)]. Show that $g_1'(t) = d$, $g_3'(u) = d^t$ and

$$g_2'(u) = \begin{pmatrix} \dfrac{\partial^2 f}{\partial x_1 \partial x_1}(u) & \cdots & \dfrac{\partial^2 f}{\partial x_1 \partial x_n}(u) \\ \vdots & \ddots & \vdots \\ \dfrac{\partial^2 f}{\partial x_n \partial x_1}(u) & \cdots & \dfrac{\partial^2 f}{\partial x_n \partial x_n}(u) \end{pmatrix}.$$

Finally prove that $f_d''(t) = d^t \nabla^2 f(x_0 + td)d$.

**Solution 9.4.** As already indicated, $f'_d(t) = \nabla f(x_0 + td)d = g_3(g_2(g_1(t)))$. After computing

$$g'_1(t) = d, \quad g'_2(u) = \nabla^2 f(u), \quad \text{and} \quad g'_3(u) = d^t,$$

it follows by Theorem 8.12 that

$$f''_d(t) = g'_3(g_2(g_1(t)))g'_2(g_1(t))g'_1(t) = d^t \nabla^2 f(x_0 + td)d.$$

★    ★    ★

**Exercise 9.5.** Suppose that $f : U \to \mathbb{R}$ is a differentiable function, where $U \subseteq \mathbb{R}^2$ is an open subset, such that the second order partial derivatives exist at a point $v \in U$. Is it true that

$$\frac{\partial^2 f}{\partial x \, \partial y}(v) = \frac{\partial^2 f}{\partial y \, \partial x}(v)?$$

**Solution 9.5.** Exercise 8.3 is an example of a function $f : \mathbb{R}^2 \to \mathbb{R}$ where the second order partial derivatives exist in $(0,0)$ but with

$$\frac{\partial^2 f}{\partial x \partial y}(0,0) = 1 \quad \text{and} \quad \frac{\partial^2 f}{\partial y \partial x}(0,0) = -1.$$

(According to Theorem 8.5, a sufficient condition for equality is that the second order partial derivatives are continuous.)

★    ★    ★

**Exercise 9.6.** Suppose that $f : U \to \mathbb{R}$ is a function differentiable in $x_0 \in U$, where $U \subseteq \mathbb{R}^n$ is an open subset.

(i) Prove for small $\lambda > 0$, that

$$f(x_0 + \lambda d) = f(x_0) + \lambda \nabla f(x_0)d + \lambda \epsilon(\lambda d)$$

according to Definition 8.1, where $d$ is a unit vector.

(ii) Now suppose that $\xi \in \mathbb{R}^n$ satisfies $f(x_0 + \lambda d) \geq f(x_0) + \lambda \xi^t d$ for every unit vector $d$ and sufficiently small $\lambda > 0$. Show that $\xi^t = \nabla f(x_0)$ using and proving the inequality

$$(\nabla f(x_0) - \xi^t)d + \epsilon(\lambda d) \geq 0$$

for $\lambda > 0$ sufficiently small. Conclude that $\partial f(x_0) = \{\nabla f(x_0)^t\}$.

**Solution 9.6.**     (i) We know that the identity in Definition 8.1 holds for $h = \lambda d$ for $\lambda > 0$ sufficiently small since $0 \in O$ and $O$ is open. After noting that $|\lambda d| = \lambda$, since $d$ is a unit vector, and $C = \nabla f(x_0)$ by Proposition 8.4, the result follows.

(ii) Using (i) together with the assumption on $\xi$ gives

$$(\nabla f(x_0) - \xi^t)d + \epsilon(\lambda d) \geq 0.$$

By letting $\lambda \to 0$ in the above we obtain

$$(\nabla f(x_0) - \xi^t)d \geq 0,$$

where we have used that $\epsilon : O \to \mathbb{R}^n$ is continuous in 0 with $\epsilon(0) = 0$. Since this inequality holds for every unit vector $d$ we find that

$$(\nabla f(x_0) - \xi^t)(\pm e_i) \geq 0,$$

where $e_i$ is the $i$-th canonical basis vector. From this we infer that $\xi^t = \nabla f(x_0)$. Any subgradient $\xi$ satisfies $f(x_0 + \lambda d) \geq f(x_0) + \lambda \xi^t d$, so it must be equal to $\nabla f(x_0)$ by the argument above. As a consequence, if $\partial f(x_0) \neq \emptyset$ (for instance if $f$ is convex by Theorem 9.3) we have $\partial f(x_0) = \{\nabla f(x_0)\}$.

$$\star \quad \star \quad \star$$

**Exercise 9.7.** Prove that a bounded convex function $f : \mathbb{R}^n \to \mathbb{R}$ is constant.

**Solution 9.7.** For any $x_0 \in \mathbb{R}^n$ there exists, by Theorem 9.3, $\xi \in \partial f(x_0)$ meaning that

$$2M \geq \xi^t(x - x_0)$$

for every $x \in \mathbb{R}^n$, where $M \in \mathbb{R}$ bounds $f$. This must necessarily imply that $\xi = 0$ (use $x = x_0 \pm k e_i$ for large $k$). By Exercise 9.2 this shows that every point is a global minimum for $f$ and thus, it is constant.

$$\star \quad \star \quad \star$$

**Exercise 9.8.** Prove that

$$f(x, y) = x^2 + y^2$$

is a strictly convex function from $\mathbb{R}^2$ to $\mathbb{R}$.

**Solution 9.8.** We compute the Hessian of $f$ as

$$\nabla^2 f(x,y) = \begin{pmatrix} 2 & 0 \\ 0 & 2 \end{pmatrix}.$$

By Lemma 9.6 $\nabla^2 f(x,y)$ is positive definite for every $(x,y) \in \mathbb{R}^2$, so we conclude that $f$ is strictly convex using Theorem 9.5.

$$\star \quad \star \quad \star$$

**Exercise 9.9.** Consider

$$f(x,y) = x^3 + y^3.$$

Give examples of two convex subsets $C_1$ and $C_2$ of $\mathbb{R}^2$, such that $f$ is a convex function on $C_1$ but not a convex function on $C_2$.

**Solution 9.9.** Theorem 9.5 gives a necessary and sufficient condition for $f$ being convex, namely that

$$\nabla^2 f(x,y) = \begin{pmatrix} 6x & 0 \\ 0 & 6y \end{pmatrix}$$

is positive semidefinite on the given open set. From Lemma 9.6 this is the case on $C_1 = (0,\infty) \times (0,\infty)$ but not on $C_2 = (-\infty,0) \times (-\infty,0)$.

$$\star \quad \star \quad \star$$

**Exercise 9.10.** Is $f(x,y) = \cos(x) + \sin(y)$ strictly convex on some nonempty open convex subset of the plane?

**Solution 9.10.** Since

$$\nabla^2 f(x,y) = \begin{pmatrix} -\cos(x) & 0 \\ 0 & -\sin(y) \end{pmatrix},$$

$f$ will be strictly convex on $C = (\frac{\pi}{2}, \pi) \times (\pi, \frac{3\pi}{2})$ by arguments as in previous exercises.

$$\star \quad \star \quad \star$$

**Exercise 9.11.** Let $f : \mathbb{R}^2 \to \mathbb{R}$ be given by

$$f(a,b) = (1 - a - b)^2 + (5 - 2a - b)^2 + (2 - 3a - b)^2.$$

(i)  Show that $f$ is a convex function. Is $f$ strictly convex?

(ii)  Find $\min\{f(a,b) \,|\, (a,b) \in \mathbb{R}^2\}$. Is this minimum unique?

(iii)  Give a geometric interpretation of the minimization problem in (ii).

**Solution 9.11.**      (i) The Hessian is

$$\nabla^2 f(a,b) = \begin{pmatrix} 28 & 12 \\ 12 & 6 \end{pmatrix}.$$

This matrix is positive definite for every $(a,b) \in \mathbb{R}^2$ by Example 9.7 implying that $f$ is strictly convex.

(ii)  The gradient vector is

$$\nabla f(a,b) = (28a + 12b - 34, 12a + 6b - 16).$$

The system of equations $\nabla f(a,b) = 0$ has the solution $(\hat{a}, \hat{b}) = (1/2, 5/3)$, and this is a global minimum for $f$ by Corollary 9.2. Uniqueness of the minimum follows by Lemma 7.6.

(iii)  By minimizing $f$ we follow the least squares method (see Example 9.8), specifically $y = \hat{a}x + \hat{b}$ is the line that minimizes the squared (vertical) deviations from the line to the points $(1,1)$, $(2,5)$, and $(3,2)$.

$$\star \quad \star \quad \star$$

**Exercise 9.12.** Show that $f : \mathbb{R}^2 \to \mathbb{R}$ given by

$$f(x,y) = \log(e^x + e^y)$$

is a convex function. Is $f$ strictly convex?

**Solution 9.12.** The Hessian of $f$ is

$$\nabla^2 f(x,y) = \frac{e^{x+y}}{(e^x + e^y)^2} \begin{pmatrix} 1 & -1 \\ -1 & 1 \end{pmatrix}.$$

By Example 9.16 $\nabla^2 f(x,y)$ is positive semidefinite for every $(x,y) \in \mathbb{R}^2$, and therefore $f$ is convex. However, the function is not strictly convex (check the definition) since $f(x,x) = \log(2) + x$ for $x \in \mathbb{R}^n$.

$$\star \quad \star \quad \star$$

**Exercise 9.13.** Can a matrix with a 0 in the upper left hand corner be positive definite? How about positive semidefinite?

**Solution 9.13.** If a matrix $A$ has 0 in the upper left hand corner then $e_1^t A e_1 = 0$, so it is not positive definite. It may happen that it is positive semidefinite; for instance, this is the case if $A = 0$.

$$\star \quad \star \quad \star$$

**Exercise 9.14.** Let us call an arbitrary $n \times n$ matrix $C$ positive semidefinite, if $v^t C v \geq 0$ for every $v \in \mathbb{R}^n$. Suppose that $A$ and $B$ are positive semidefinite $n \times n$ matrices. Is the (matrix) product $AB$ positive semidefinite? Suppose that

$$A = \begin{pmatrix} a_{11} & a_{12} \\ a_{12} & a_{22} \end{pmatrix} \quad \text{and} \quad B = \begin{pmatrix} b_{11} & b_{12} \\ b_{12} & b_{22} \end{pmatrix}$$

are symmetric and positive semidefinite. Is

$$\begin{pmatrix} a_{11} b_{11} & a_{12} b_{12} \\ a_{12} b_{12} & a_{22} b_{22} \end{pmatrix}$$

positive semidefinite? (See also Exercise 9.19.)

**Solution 9.14.** Since

$$\begin{pmatrix} 1 & 1 \\ 1 & 1 \end{pmatrix} \begin{pmatrix} 1 & -2 \\ -2 & 4 \end{pmatrix} = \begin{pmatrix} -1 & 2 \\ -1 & 2 \end{pmatrix}$$

we see that the matrix product of two positive semidefinite matrices is not necessarily positive semidefinite. The matrix

$$\begin{pmatrix} a_{11} b_{11} & a_{12} b_{12} \\ a_{12} b_{12} & a_{22} b_{22} \end{pmatrix}$$

will be positive semidefinite. To see this, use Example 9.16 to get $a_{11} a_{22} \geq a_{12}^2$, $b_{11} b_{22} \geq b_{12}^2$, and $a_{11}, a_{22}, b_{11}, b_{22} \geq 0$. From these inequalities it follows that $a_{11} b_{11}, a_{22} b_{22} \geq 0$ and $a_{11} b_{11} a_{22} b_{22} - a_{12}^2 b_{12}^2 \geq 0$, and then the same example ensures that the matrix is positive semidefinite.

$$\star \quad \star \quad \star$$

**Exercise 9.15.** Let $A$ be a positive semidefinite $n \times n$ matrix. Suppose that $v^t A v = 0$. Prove that $Av = 0$.

**Solution 9.15.** By Corollary 9.18 there exists a matrix $S$ such that $A = S^t S$. Consequently, $0 = v^t A v = |Sv|^2$ implying $Sv = 0$ and thus, $Av = 0$.

$$\star \quad \star \quad \star$$

**Exercise 9.16.** Consider the real symmetric matrix

$$A = \begin{pmatrix} a & c \\ c & b \end{pmatrix}.$$

Prove from scratch that $A$ has a real eigenvalue (without using Theorem 9.24).

**Solution 9.16.** We need to argue that there exists $\lambda \in \mathbb{R}$ such that $\det(A - \lambda I) = (a - \lambda)(b - \lambda) - c^2 = 0$. This holds if and only if $(a + b)^2 - 4(ab - c^2) = 4c^2 + (a - b)^2 \geq 0$, which is indeed the case.

$$\star \quad \star \quad \star$$

**Exercise 9.17.** Let

$$T = \begin{pmatrix} \cos\theta & -\sin\theta \\ \sin\theta & \cos\theta \end{pmatrix},$$

where $\theta \in \mathbb{R}$. Prove that $T^{-1} = T^t$.

**Solution 9.17.** After noting that

$$TT^t = \begin{pmatrix} \cos\theta & -\sin\theta \\ \sin\theta & \cos\theta \end{pmatrix} \begin{pmatrix} \cos\theta & \sin\theta \\ -\sin\theta & \cos\theta \end{pmatrix} = I$$

the result is shown.

$$\star \quad \star \quad \star$$

**Exercise 9.18.** Consider the optimization problem

$$M = \max x^t A x$$
$$x \in S^n,$$

in the proof of Theorem 9.24. Suppose that $\lambda = M$ and $z^t A z = \lambda$ with $z^t z = 1$. Prove for $B = A - \lambda I$ that

(i) $z^t B z = 0$
(ii) $x^t B x \leq 0$ for every $x \in \mathbb{R}^n$.

Use this to prove $Bz = 0$ by applying that

$$(z + tBz)^t B(z + tBz) \leq 0$$

for every $t \in \mathbb{R}$. (This gives a formal alternative to the use of Lagrange multipliers in the proof of Theorem 9.24.)

**Solution 9.18.**    (i) Note that $z^t B z = z^t A z - \lambda z^t z = 0$.

(ii)  For $x \in \mathbb{R}^n \setminus \{0\}$ we let $u = x/|x|$ and notice that $x^t B x = |x|^2 u^t B u$. Since $u \in S^n$ it follows that $u^t B u \leq z^t B z = 0$ and thus, $x^t B x \leq 0$.

For any given $t \in \mathbb{R}$, (ii) gives

$$0 \geq (z + tBz)^t B(z + tBz) = t \left( 2|Bz|^2 + t(Bz)^t B(Bz) \right)$$

where we have used the fact that $B$ is symmetric. If $Bz \neq 0$ then $|Bz| > 0$ and $2|Bz|^2 > t|(Bz)^t B(Bz)|$ for $t > 0$ sufficiently small and in that case, the inequality above is violated. Hence, we conclude that $Bz = 0$.

⋆   ⋆   ⋆

**Exercise 9.19.** Show that any matrix of the form $C^t C$, where $C$ is an $m \times n$ matrix is positive semidefinite. If $v \in \mathbb{R}^n$ show that $vv^t$ is a positive semidefinite matrix of rank one.

**Solution 9.19.** For any $x \in \mathbb{R}^n$ it holds that $x^t C^t C x = |Cx|^2 \geq 0$ showing that $C$ is positive semidefinite. Particularly, $vv^t$ is positive semidefinite for $v \in \mathbb{R}^n$. Writing $vv^t = \begin{pmatrix} v_1 v & \cdots & v_n v \end{pmatrix}$ emphasizes that the columns are all multiples of $v$ meaning that $vv^t$ has rank one for $v \neq 0$.

⋆   ⋆   ⋆

**Exercise 9.20.** The *Hadamard product* $A \circ B$ of two matrices $A$ and $B$ with the same dimensions is given by $(A \circ B)_{ij} = A_{ij} B_{ij}$. Prove that $A \circ B$ is positive semidefinite if $A$ and $B$ are two positive semidefinite $n \times n$ matrices. Hint: use Exercise 9.19! First prove this, when $B$ has rank one and can be written $B = vv^t$ using that

$$A \circ vv^t = D^t A D,$$

where

$$D = \begin{pmatrix} v_1 & 0 & \cdots & 0 \\ 0 & v_2 & \cdots & 0 \\ \vdots & \vdots & \ddots & \vdots \\ 0 & 0 & \cdots & v_n \end{pmatrix}$$

and $v = (v_1, \ldots, v_n)^t$. Then move on to the general case, where $B$ is a sum of rank one positive semidefinite matrices.

**Solution 9.20.** First consider the Hadamard product $A \circ vv^t$ for some $v \in \mathbb{R}^n$. We find

$$(A \circ vv^t)_{ij} = A_{ij}v_iv_j = (D^tAD)_{ij}$$

where $D$ is given above. Since $A$ is positive semidefinite, Corollary 9.18 gives the existence of $S$ such that $A = S^tS$. Writing $D^tAD = (SD)^tSD$ together with Exercise 9.19 shows that $A \circ vv^t$ is positive semidefinite. Consider a general positive semidefinite $n \times n$ matrix $B$. Again we use Corollary 9.18 to write

$$
\begin{aligned}
A \circ B &= A \circ (v_1v_1^t + v_2v_2^t + \cdots + v_nv_n^t) \\
&= A \circ v_1v_1^t + A \circ v_2v_2^t + \cdots + A \circ v_nv_n^t
\end{aligned}
\tag{9.4}
$$

where $v_i \in \mathbb{R}^n$. Here we have used the distributive property of the Hadamard product which is not too hard to verify. This shows that (9.4) is a sum of positive semidefinite matrices and hence, positive semidefinite itself.

$$\star \quad \star \quad \star$$

**Exercise 9.21.** Let

$$S = \begin{pmatrix} A & 0 \\ 0 & B \end{pmatrix}$$

denote a symmetric $(r + s) \times (r + s)$ matrix, where $A$ is a symmetric $r \times r$ matrix and $B$ a symmetric $s \times s$ matrix. Prove that $S$ is positive definite if and only if $A$ and $B$ are positive definite. Prove also the same statement with definite replaced by semidefinite.

**Solution 9.21.** Since the arguments are similar, we only show the case where $A$ and $B$ are positive definite matrices. For any $(x, y) \in \mathbb{R}^{r+s}$ we have

$$(x, y)S \begin{pmatrix} x \\ y \end{pmatrix} = x^tAx + y^tBy.
\tag{9.5}$$

Assume $S$ is positive definite. If $x \neq 0$, we can use (9.5) on $(x, 0)$ to find that $x^tAx > 0$ meaning that $A$ is positive definite. Similarly we get that $B$ is positive definite. Now suppose that $A$ and $B$ are positive definite. Then if $(x, y) \neq 0$, we can assume $x \neq 0$ such that $x^tAx > 0$. Since we always have $y^tBy \geq 0$, (9.5) shows that $S$ is positive definite.

$$\star \quad \star \quad \star$$

**Exercise 9.22.** Give an example of a non-symmetric $n \times n$ matrix $A$, such that $x^t A x \geq 0$ for every $x \in \mathbb{R}^n$.

**Solution 9.22.** Consider the matrix $A$ with entries given by

$$a_{ij} = \begin{cases} 1 & \text{if } i = j \\ 2 & \text{if } i = n, \ j = 1 \\ 0 & \text{otherwise} \end{cases}$$

that is, the matrix with ones on the diagonal, two in the lower left corner, and zeros everywhere else. Then we find that

$$x^t A x = x_1^2 + x_2^2 + \cdots + x_n^2 + 2x_1 x_n = (x_1 + x_n)^2 + x_2^2 + \cdots + x_{n-1}^2 \geq 0$$

for every $x \in \mathbb{R}^n$.

$$\star \quad \star \quad \star$$

**Exercise 9.23.** Compute the set

$$\left\{ (a, b) \in \mathbb{R}^2 \ \middle|\ \begin{pmatrix} 2 & 1 & a \\ 1 & 1 & 1 \\ a & 1 & b \end{pmatrix} \text{ is positive definite} \right\}.$$

Same question with positive semidefinite. Sketch and compare the two subsets of the plane.

**Solution 9.23.** Theorem 9.11 says that

$$A = \begin{pmatrix} 2 & 1 & a \\ 1 & 1 & 1 \\ a & 1 & b \end{pmatrix}$$

is positive definite if and only if its leading principal minors are positive giving us that

$$\{(a, b) \in \mathbb{R}^2 \mid A \text{ is positive definite}\} = \{(a, b) \in \mathbb{R}^2 \mid 2a - a^2 + b - 2 > 0\}.$$

If, on the other hand, we wish to find $a, b \in \mathbb{R}$ such that $A$ is positive semidefinite Theorem 9.15 says that we must have

$$\begin{aligned} 2a - a^2 + b - 2 &\geq 0 \\ b - 1 &\geq 0 \\ b &\geq 0 \\ -a^2 + 2b &\geq 0. \end{aligned}$$

Only the first inequality is relevant; e.g. writing

$$b \geq a^2 - 2a + 2 = \tfrac{1}{2}a^2 + \left(\tfrac{1}{\sqrt{2}}a - \sqrt{2}\right)^2 \geq \tfrac{1}{2}a^2$$

ensures that the last inequality is satisfied if the first is so. Consequently, we find that

$$\{(a, b) \in \mathbb{R}^2 \mid A \text{ is positive semidefinite}\} = \{(a, b) \in \mathbb{R}^2 \mid 2a - a^2 + b - 2 \geq 0\}.$$

In Figure 9.1 is sketched the sets above. Note that the first set does not contain the boundary but the second does.

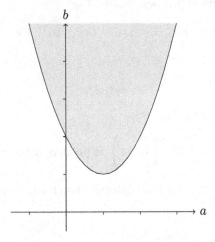

**Figure 9.1:** The set of $(a, b)$ such that $A$ is positive definite (semidefinite) is shaded in grey without (with) the boundary.

$$\star \quad \star \quad \star$$

**Exercise 9.24.** Let $f : \mathbb{R}^2 \to \mathbb{R}$ be given by

$$f(x, y) = ax^2 + by^2 + cxy,$$

where $a, b, c \in \mathbb{R}$.

(i) Show that $f$ is a convex function if and only if $a \geq 0, b \geq 0$ and $4ab - c^2 \geq 0$.

(ii) Suppose now that $a > 0$ and $4ab - c^2 > 0$. Show that $g(x, y) = f(x, y) + x + y$ has a unique global minimum and give a formula for this minimum in terms of $a, b$ and $c$.

In the following we will assume that $f$ is a convex function.

(iii) Show that $f$ is not strictly convex if $a = 0$.

(iv) Show that $f$ is not strictly convex if $a > 0$ and $4ab - c^2 = 0$.

**Solution 9.24.**   (i) We may (as in Example 9.26) write

$$f(x,y) = (x,y)A\begin{pmatrix} x \\ y \end{pmatrix}, \quad \text{where} \quad A = \begin{pmatrix} a & c/2 \\ c/2 & b \end{pmatrix}.$$

Now Theorem 9.28 gives that $f$ is convex if and only if $A$ is positive semidefinite which according to Theorem 9.15 is equivalent to

$$a \geq 0, \quad b \geq 0, \quad \text{and} \quad 4ab - c^2 \geq 0.$$

(ii)   Since $\nabla^2 g(x,y) = 2A$ then $g$ is strictly convex by Theorem 9.5. Corollary 9.2 gives that a global minimum is characterized by the condition $\nabla g(x,y) = 0$ which is solved by

$$\begin{pmatrix} x \\ y \end{pmatrix} = \frac{1}{4ab - c^2}\begin{pmatrix} c - 2b \\ c - 2a \end{pmatrix}.$$

Lemma 7.6 ensures the uniqueness of the minimum.

(iii)-(iv) This is a consequence of Theorem 9.28. This result should be compared with Theorem 9.5 since our conclusion heavily relies on $f$ being a quadratic form.

$$\star \quad \star \quad \star$$

**Exercise 9.25.**   Recall that a convex cone in a euclidean vector space $\mathbb{R}^n$ is a subset $C \subseteq \mathbb{R}^n$ such that $x + y \in C$ and $\lambda x \in C$ for every $x, y \in C$ and every $\lambda \geq 0$. Let

$$S = \left\{ \begin{pmatrix} a & b & c \\ b & d & e \\ c & e & f \end{pmatrix} \,\middle|\, a, b, c, d, e, f \in \mathbb{R} \right\}$$

be the set of symmetric $3 \times 3$ matrices. Notice that $S$ may be viewed as the vector space $\mathbb{R}^6 = \{(a, b, c, d, e, f) \,|\, a, b, c, d, e, f \in \mathbb{R}\}$.

(i) Let $x \in \mathbb{R}^3$. Show that

$$C_x = \{A \in S \,|\, x^t Ax \geq 0\}$$

is a closed convex cone in $S$.

(ii) Let $\mathcal{P} \subseteq \mathcal{S}$ be the set of positive semidefinite matrices. Show that

$$\mathcal{P} = \bigcap_{x \in \mathbb{R}^n} C_x.$$

Use this to prove that $\mathcal{P}$ is a closed convex cone in $\mathcal{S}$.

(iii) Prove that the closure of the set of positive definite matrices inside $\mathcal{S}$ coincides with $\mathcal{P}$.

(iv) Let

$$S' = \left\{ \begin{pmatrix} a & b & c \\ b & d & e \\ c & e & f \end{pmatrix} \in \mathcal{P} \ \middle| \ a - b + c = 0, \ e = 0 \right\}.$$

Show that $S'$ is a convex set in $\mathcal{S}$.

**Solution 9.25.**     (i) Let $A_1, A_2 \in C_x$ and $\lambda \geq 0$, and write

$$x^t(\lambda A_1 + A_2)x = \lambda x^t A_1 x + x^t A_2 x \geq 0$$

which shows that $C_x$ is a convex cone. Furthermore, let $x^t A_n x \geq 0$ for every $n \in \mathbb{N}$ and $A_n \to A$ (in the sense of usual convergence in $\mathbb{R}^6$). Then $x^t A_n x \to x^t A x$ by continuity and therefore, $x^t A x \geq 0$ which shows that $C_x$ is closed.

(ii) By the definition we have $A \in \mathcal{P}$ if and only if $A \in C_x$ for all $x$ showing the equality. Additionally, $\mathcal{P}$ must be a closed convex cone since it is an intersection of closed convex cones.

(iii) Let $\mathcal{P}'$ be the set of positive definite matrices inside $\mathcal{S}$. Since $\mathcal{P}' \subseteq \mathcal{P}$ and $\mathcal{P}$ is closed, we have $\overline{\mathcal{P}'} \subseteq \mathcal{P}$. On the other hand, consider the sequence $(A + \frac{1}{n}I)$ where $A \in \mathcal{P}$. Then

$$x^t \left( A + \tfrac{1}{n}I \right) x = x^t A x + \tfrac{1}{n}|x|^2 > 0$$

for every $x \in \mathbb{R}^n \setminus \{0\}$. Thus, $A_n \in \mathcal{P}'$ for every $n$ and $A_n \to A$ which shows that $\mathcal{P} \subseteq \overline{\mathcal{P}'}$.

(iv) Since $S'$ is the intersection of the convex set $\mathcal{P}$ and two affine hyperplanes, it is convex.

**Exercise 9.26.** Why is the subset given by the inequalities

$$x \geq 0$$
$$y \geq 0$$
$$xy - z^2 \geq 0$$

a convex subset of $\mathbb{R}^3$?

**Solution 9.26.** From Example 9.16 we know that

$$\mathcal{P} = \{(x, y, z) \in \mathbb{R}^3 \mid x \geq 0, \, y \geq 0 \, xy - z^2 \geq 0\}$$

$$= \left\{(x, y, z) \in \mathbb{R}^3 \mid \begin{pmatrix} x & z \\ z & y \end{pmatrix} \text{ is positive semidefinite}\right\}.$$

To verify that this is a convex set, argue as in Exercise 9.25(i).

$\star \quad \star \quad \star$

**Exercise 9.27.** Let $\mathcal{S}$ denote the convex cone of positive semidefinite $n \times n$ matrices in the vector space of symmetric $n \times n$ matrices. Prove that

$$\{vv^t \mid v \in \mathbb{R}^n \setminus \{0\}\}$$

are the extreme rays in $\mathcal{S}$ (use Exercise 9.15 and Corollary 9.18).

**Solution 9.27.** Consider $v \in \mathbb{R}^n \setminus \{0\}$. We start by showing that $vv^t$ is an extreme ray in $\mathcal{S}$. Suppose that $A_1, A_2 \in \mathcal{S}$ with $vv^t = A_1 + A_2$. As a consequence of this,

$$0 \leq x^t A_1 x \leq x^t (vv^t) x \tag{9.6}$$

for every $x \in \mathbb{R}^n$, from which it appears (when using Exercise 9.15) that $\mathcal{N}(vv^t) \subseteq \mathcal{N}(A_1)$ where $\mathcal{N}(A) := \{x \in \mathbb{R}^n \mid Ax = 0\}$. Consequently, when using that rk $vv^t = 1$ and $\dim \mathcal{N}(A) + \text{rk } A = n$ (see Exercise 2.3), we find that rk $A_1 \leq 1$. If rk $A_1 = 0$ then $A_1 = 0$ and thus, $A_1 \in \text{cone}(\{vv^t\})$. On the other hand, by considering the proof of Corollary 9.18 it is evident that there exists $u \in \mathbb{R}^n \setminus \{0\}$ such that $A_1 = uu^t$ when rk $A_1 = 1$. If $u$ and $v$ are linearly independent it follows from Theorem B.9 that we may find $x \in \mathbb{R}^n$ with $u^t x \neq 0$ and $v^t x = 0$. However, this is a contradiction since

$$(u^t x)^2 = x^t A_1 x \leq x^t (vv^t) x = (v^t x)^2$$

by (9.6). Therefore there exists $\alpha$ with $uu^t = \alpha^2 vv^t$ and consequently, $A_1 \in \text{cone}(\{vv^t\})$. The same arguments show that $A_2 \in \text{cone}(\{vv^t\})$ and by the characterization in [U.C., (5.2)], $vv^t$ is an extreme ray in $\mathcal{S}$.

For any $A \in \mathcal{S}$, Corollary 9.18 gives that

$$A = v_1 v_1^t + (v_2 v_2^t + \cdots + v_d v_d^t)$$

where $d \leq n$ and $v_i v_i^t \neq 0$ for $1 \leq i \leq d$. If $A$ is an extreme ray in $S$ there exists $\lambda \geq 0$ such that $\lambda A = v_1 v_1^t$, and since $v_1 v_1^t \neq 0$ we must have that $\lambda > 0$. Defining $v = \lambda^{-1/2} v_1$ yields $A = v v^t$ which was to be shown.

$$\star \quad \star \quad \star$$

**Exercise 9.28.** Let

$$f(x, y, z) = 3x^2 - 2xy - 2xz + 3y^2 + 3z^2 - 2yz.$$

(i) Show that $f(x, y, z) \geq 0$ for every $x, y, z \in \mathbb{R}$.
(ii) Show that $f(x, y, z) = 0$ if and only if $x = 0$, $y = 0$ and $z = 0$.

**Solution 9.28.** Note that

$$f(x, y, z) = (x, y, z) \begin{pmatrix} 3 & -1 & -1 \\ -1 & 3 & -1 \\ -1 & -1 & 3 \end{pmatrix} \begin{pmatrix} x \\ y \\ z \end{pmatrix}.$$

We verify that the matrix is positive definite by computing its leading principal minors, which answers (i) and (ii).

$$\star \quad \star \quad \star$$

**Exercise 9.29.** Let $f_{(a,b)}$ be given by

$$f_{(a,b)}(x, y, z) = ax^2 + by^2 + 4z^2 + 2xy + 4xz + 6yz$$

and let $S$ denote the set of $(a, b)$ such that $f_{(a,b)}$ is strictly convex. Show that $S$ is non-empty and convex. Give a finite set of inequalities that defines $S$.

**Solution 9.29.** For a given point $(a, b) \in \mathbb{R}^2$ we have that

$$f_{(a,b)}(x, y, z) = (x, y, z) \begin{pmatrix} a & 1 & 2 \\ 1 & b & 3 \\ 2 & 3 & 4 \end{pmatrix} \begin{pmatrix} x \\ y \\ z \end{pmatrix}.$$

We know from Theorem 9.28(1) that $f_{(a,b)}$ is strictly convex if and only if the leading principal minors of the matrix associated with $f_{(a,b)}$ are positive, and therefore we obtain

$$S = \left\{ \begin{pmatrix} a \\ b \end{pmatrix} \in \mathbb{R}^2 \;\middle|\; \begin{array}{rcl} a & > & 0 \\ ab & > & 1 \\ -9a - 4b + 4ab & > & -8 \end{array} \right\}.$$

From this it appears that $S$ is non-empty as $(2, 4) \in S$. It should not be too difficult to verify that $(a, b) \mapsto f_{(a,b)}(x, y, z)$ for fixed $(x, y, z) \in \mathbb{R}^3$ is an affine map. Therefore, it follows that $f_{((1-\lambda)a_1+\lambda a_2,(1-\lambda)b_1+\lambda b_2)}$ is strictly convex if $f_{(a_1,b_1)}$ and $f_{(a_2,b_2)}$ are strictly convex and $0 \le \lambda \le 1$ meaning that $S$ is convex.

$$\star \quad \star \quad \star$$

**Exercise 9.30.** Let

$$f(x, y) = 5x^2 - 2xy + 5y^2.$$

Based on Example 9.25, prove that $f(x, y) \ge 0$ using an orthonormal basis of eigenvectors for a symmetric matrix associated with $f$.

**Solution 9.30.** First, we write $f$ as

$$f(x, y) = (x, y) \begin{pmatrix} 5 & -1 \\ -1 & 5 \end{pmatrix} \begin{pmatrix} x \\ y \end{pmatrix}.$$

Letting $A$ denote the matrix that we associate with $f$,

$$\det(A - \lambda I) = (5 - \lambda)^2 - 1 = 0$$

for $\lambda = 4$ or $\lambda = 6$. We find that

$$(A - 4I)u = \begin{pmatrix} 1 & -1 \\ -1 & 1 \end{pmatrix} u = 0$$

for $u = (1, 1)/\sqrt{2}$ and similarly,

$$(A - 6I)v = \begin{pmatrix} -1 & -1 \\ -1 & -1 \end{pmatrix} v = 0$$

for $v = (-1, 1)/\sqrt{2}$. Using this orthonormal basis of eigenvectors for $A$ we define

$$T = \begin{pmatrix} \frac{1}{\sqrt{2}} & -\frac{1}{\sqrt{2}} \\ \frac{1}{\sqrt{2}} & \frac{1}{\sqrt{2}} \end{pmatrix}.$$

Since

$$T^t A T = \begin{pmatrix} 4 & 0 \\ 0 & 6 \end{pmatrix}$$

it follows directly from Lemma 9.6 that $A$ is positive definite and thus, $f(x, y) > 0$ whenever $(x, y) \ne 0$.

$$\star \quad \star \quad \star$$

**Exercise 9.31.** Let

$$f(x,y) = 2x^2 + 3y^2 + 4xy.$$

Prove that $f(x,y) \geq 0$ for every $x, y \in \mathbb{R}$. Plot the points $(x,y) \in \mathbb{R}^2$ with $f(x,y) = 1$.

**Solution 9.31.** We write $f$ as

$$f(x,y) = (x,y) \begin{pmatrix} 2 & 2 \\ 2 & 3 \end{pmatrix} \begin{pmatrix} x \\ y \end{pmatrix}.$$

Since we are going to plot $f(x,y) = 1$ we will, instead of checking minors of the matrix associated with $f$, observe that (see Example 9.7)

$$\begin{pmatrix} 1 & 0 \\ -1 & 1 \end{pmatrix} \begin{pmatrix} 2 & 2 \\ 2 & 3 \end{pmatrix} \begin{pmatrix} 1 & -1 \\ 0 & 1 \end{pmatrix} = \begin{pmatrix} 2 & 0 \\ 0 & 1 \end{pmatrix}. \tag{9.7}$$

Now it follows from Lemma 9.6 that the matrix, and therefore $f$ itself, is positive definite. Furthermore, (9.7) shows that

$$\begin{pmatrix} 2 & 2 \\ 2 & 3 \end{pmatrix} = \begin{pmatrix} 1 & 0 \\ 1 & 1 \end{pmatrix} \begin{pmatrix} 2 & 0 \\ 0 & 1 \end{pmatrix} \begin{pmatrix} 1 & 1 \\ 0 & 1 \end{pmatrix}.$$

This means that

$$f(x,y) = (x+y,y) \begin{pmatrix} 2 & 0 \\ 0 & 1 \end{pmatrix} \begin{pmatrix} x+y \\ y \end{pmatrix} = 2(x+y)^2 + y^2.$$

Therefore $f(x,y) = 1$ if and only if $(x,y) = \left( \pm\sqrt{\frac{1}{2}(1 - t^2)} - t, t \right)$ for $0 \leq t \leq 1$, and this simplifies the task of plotting $f$ (which is done in Figure 9.2).

$$\star \quad \star \quad \star$$

**Exercise 9.32.** Is the quadratic form

$$f(x,y,z) = 2x^2 + 2y^2 + 4z^2 + 2xy + 4xz + 6yz$$

positive semidefinite?

**Solution 9.32.** One can check that $f\left(0, -2, \frac{3}{2}\right) = -1 < 0$ which means that $f$ is not positive semidefinite. Another way to reach the same conclusion is by noting that if $f$ should be positive semidefinite, the three inequalities in Exercise 9.29 should (at least weakly) be satisfied with $(a,b) = (2,2)$ cf. Theorem 9.15.

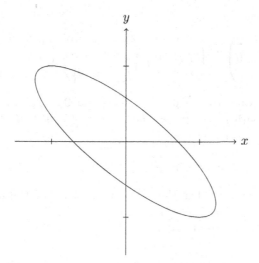

**Figure 9.2:** The points of $\mathbb{R}^2$ such that $2x^2 + 3y^2 + 4xy = 1$.

$$\star \quad \star \quad \star$$

**Exercise 9.33.** Let

$$f(x, y, z) = x^2 + 3y^2 + 5z^2 + 4xz + 10xy.$$

Compute a basis $b_1, b_2, b_3 \in \mathbb{R}^3$, such that

$$f(v) = \lambda_1(b_1^t v)^2 + \lambda_2(b_2^t v)^2 + \lambda_3(b_3^t v)^2$$

for $v = (x, y, z)^t$. Compute $p$ and $q$ as given in Theorem 9.28 for $f$. Find $x, y, z \in \mathbb{R}$ with $f(x, y, z) < 0$.

[Extra credit!] Use a computer to sketch the set

$$\{(x, y, z) \in \mathbb{R}^3 \mid f(x, y, z) = 1\}.$$

**Solution 9.33.** First write

$$f(x, y, z) = (x, y, z) \begin{pmatrix} 1 & 5 & 2 \\ 5 & 3 & 0 \\ 2 & 0 & 5 \end{pmatrix} \begin{pmatrix} x \\ y \\ z \end{pmatrix}.$$

Using the procedure as in the proof of Theorem 9.17, we find

$$\begin{pmatrix} 1 & 0 & 0 \\ -5 & 1 & 0 \\ \frac{3}{11} & -\frac{5}{11} & 1 \end{pmatrix} \begin{pmatrix} 1 & 5 & 2 \\ 5 & 3 & 0 \\ 2 & 0 & 5 \end{pmatrix} \begin{pmatrix} 1 & -5 & \frac{3}{11} \\ 0 & 1 & -\frac{5}{11} \\ 0 & 0 & 1 \end{pmatrix} = \begin{pmatrix} 1 & 0 & 0 \\ 0 & -22 & 0 \\ 0 & 0 & \frac{61}{11} \end{pmatrix}$$

which means that

$$\begin{pmatrix} 1 & 5 & 2 \\ 5 & 3 & 0 \\ 2 & 0 & 5 \end{pmatrix} = \begin{pmatrix} 1 & 0 & 0 \\ 5 & 1 & 0 \\ 2 & \frac{5}{11} & 1 \end{pmatrix} \begin{pmatrix} 1 & 0 & 0 \\ 0 & -22 & 0 \\ 0 & 0 & \frac{61}{11} \end{pmatrix} \begin{pmatrix} 1 & 5 & 2 \\ 0 & 1 & \frac{5}{11} \\ 0 & 0 & 1 \end{pmatrix}.$$

Therefore, according to the proof of Theorem 9.28(2), we have found a representation with $b_1 = (1, 5, 2)$, $b_2 = \left(0, 1, \frac{5}{11}\right)$, $b_3 = (0, 0, 1)$, $\lambda_1 = 1$, $\lambda_2 = -22$, and $\lambda_3 = 61/11$ that is,

$$f(x, y, z) = (x + 5y + 2z)^2 - 22\left(y + \tfrac{5}{11}z\right)^2 + \tfrac{61}{11}z^2. \tag{9.8}$$

Therefore $p = 2$ and $q = 1$ in the notation of Theorem 9.28(3). From (9.8) we see that, for instance, $f(-5, 1, 0) < 0$.

Similar to Exercise 9.31, we note that $f(x, y, z) = 1$ if and only if

$$(x, y, z) = \left(\pm\sqrt{1 - \tfrac{61}{11}z^2 + 22\left(y + \tfrac{5}{11}z\right)^2} - 5y - 2z, y, z\right)$$

and $\frac{61}{11}z^2 - 22\left(y + \frac{5}{11}z\right)^2 \leq 1$. The set of such $(x, y, z)$'s is plotted in Figure 9.3.[1]

$$\star \quad \star \quad \star$$

**Exercise 9.34.** Let $f_a : \mathbb{R}^2 \to \mathbb{R}$ denote the quadratic form given by

$$f_a(x, y, z) = x^2 + y^2 + z^2 + 2axy + 2xz + 2yz,$$

where $a \in \mathbb{R}$.

(i) Write down the symmetric matrix associated to $f_a$.
(ii) Show that $f_a$ is not positive definite for $a = 2$.
(iii) Show that $f_a$ is not positive definite for any $a \in \mathbb{R}$.
(iv) Is there an $a \in \mathbb{R}$ such that $f_a$ is positive semidefinite?
(v) Rewrite $f_a$ for $a = 1$ as a sum of squares of linear forms in $x, y$ and $z$.

**Solution 9.34.** (i) The symmetric matrix associated to $f_a$ is

$$A_a = \begin{pmatrix} 1 & a & 1 \\ a & 1 & 1 \\ 1 & 1 & 1 \end{pmatrix}.$$

---

[1]This plot is produced in Mathematica 10.0 using the command: ContourPlot3D[(x + 5 y + 2 z)^2 - 22 (y + 5/11 z)^2 + 61/11 z^2 == 1, {x, -1, 1}, {y, -1, 1}, {z, -1, 1}].

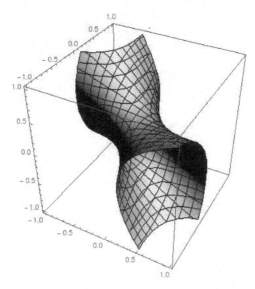

**Figure 9.3:** The set $\{(x,\,y,\,z) \in \mathbb{R}^3 \,|\, f(x,y,z) = 1\}$ in the notation of Exercise 9.33.

(ii) We find that $\det(A_2[[2],[2]]) = -3$, and Theorem 9.11 ensures that $f_2$ is not positive definite.

(iii) Once again by Theorem 9.11, $f_a$ cannot be positive definite since $\det(A_a) = -(a-1)^2$.

(iv) From (ii), $f_1$ is the only candidate to be positive semidefinte. It is not too difficult to verify that $A_1$ is in fact positive semidefinite using Example 9.16.

(v) Following the procedure from Example 9.19, we find

$$\begin{pmatrix} 1 & 1 & 1 \\ 1 & 1 & 1 \\ 1 & 1 & 1 \end{pmatrix} = \begin{pmatrix} 1 & 0 & 0 \\ 1 & 1 & 0 \\ 1 & 0 & 1 \end{pmatrix} \begin{pmatrix} 1 & 0 & 0 \\ 0 & 0 & 0 \\ 0 & 0 & 0 \end{pmatrix} \begin{pmatrix} 1 & 1 & 1 \\ 0 & 1 & 0 \\ 0 & 0 & 1 \end{pmatrix}.$$

By using this expression (see the proof of Theorem 9.28(2)) we find that $f(x,y,z) = (x+y+z)^2$.

# Chapter 10

# Convex optimization

## 10.1 Introduction

In applied mathematics, solving optimization problems is of great importance and the material covered in this chapter has been applied in various fields. The KKT conditions (Definition 10.4) have found applications in fields from economics to computer science, and it is likely that these conditions will appear several times in any career containing a branch of applied mathematics.

When encountering a convex optimization problem it is not always apparent how to 'take the first step'. As a consequence, the following solutions are often accompanied by different illustrations and geometric interpretations to help getting the first idea of how to solve the problem. The hope is that these will give a better understanding of how one could verify the KKT conditions. Not all the exercises are concerned with the KKT conditions as we touch two other aspects of convex optimization, namely the dual problem and the understanding and implementation of an interior point algorithm.

The following results give several (necessary and sufficient) optimality conditions, which will be useful when solving the optimization problems we are faced in the exercises.

**Proposition 10.1.** *Let $S \subseteq \mathbb{R}^n$ be a convex subset and $f : U \to \mathbb{R}$ a differentiable function defined on an open subset $U \supseteq S$. If $x_0 \in S$ is an optimal solution of [U.C., (10.1)], then*

$$\nabla f(x_0)(x - x_0) \geq 0 \qquad \text{for every } x \in S. \tag{10.1}$$

*If $f$ in addition is a convex function, then (10.1) implies that $x_0$ is optimal.*

147

**Definition 10.4.** The system

$$\lambda_i \geq 0$$

$$g_i(x_0) \leq 0 \quad \text{and} \quad \lambda_i g_i(x_0) = 0$$

$$\nabla f(x_0) + \lambda_1 \nabla g_1(x_0) + \cdots + \lambda_m \nabla g_m(x_0) = 0,$$

of inequalities for $i = 1, \ldots, m$ in the unknowns $x_0 \in \mathbb{R}^n$ and $\lambda_1, \ldots, \lambda_m \in \mathbb{R}$ are called the *Karush-Kuhn-Tucker (KKT) conditions* associated with the optimization problem [(U.C), (10.4)].

In the solutions we will refer to latter KKT condition as the gradient condition.

**Theorem 10.6.** *Consider the optimization problem*

$$\min\{f(x) \mid x \in S\} \tag{10.2}$$

*where $S = \{x \in \mathbb{R}^n \mid g_1(x) \leq 0, \ldots, g_m(x) \leq 0\}$ with $g_1, \ldots, g_m : \mathbb{R}^n \to \mathbb{R}$ differentiable functions and $f : U \to \mathbb{R}$ differentiable on an open set $U$ containing $S$. Let $x_0 \in S$ denote an optimal solution of (10.2).*

(1) *If*

$$\nabla g_1(x_0), \ldots, \nabla g_m(x_0)$$

*are linearly independent, then the KKT conditions are satisfied at $x_0$ for suitable $\lambda_1, \ldots, \lambda_m$.*

(2) *If $g_1, \ldots, g_m$ are convex and (10.2) is strictly feasible, then the KKT conditions are satisfied at $x_0$ for suitable $\lambda_1, \ldots, \lambda_m$.*

(3) *If $g_1, \ldots, g_m$ are affine functions i.e., $S = \{x \in \mathbb{R}^n \mid Ax \leq b\}$ for an $m \times n$ matrix $A$ and $b \in \mathbb{R}^m$, then the KKT conditions are satisfied at $x_0$ for suitable $\lambda_1, \ldots, \lambda_m$.*

(4) *If $f$ and $g_1, \ldots, g_m$ are all convex functions and the KKT conditions hold at $z$ for some $\lambda_1, \ldots, \lambda_m \in \mathbb{R}$, then $z$ is an optimal solution of (10.2).*

**Theorem 10.15.** *Let $V = \{x_1, \ldots, x_m\} \subseteq \mathbb{R}^n$ and*

$$P = \operatorname{conv}(\{x_1, \ldots, x_m\})$$

$$= \{\lambda_1 x_1 + \cdots + \lambda_m x_m \mid \lambda_i \geq 0, \ \lambda_1 + \cdots + \lambda_m = 1\}.$$

*The optimization problem*

$$\max\{f(x) \,|\, x \in P\}$$

*has a global maximum at $x_0 \in V$, where*

$$f(x_0) = \max\{f(x_1), \ldots, f(x_m)\}.$$

*This maximum is unique if and only if $f(x_0) = f(x_j)$ for a unique $j = 1, \ldots, m$.*

## 10.2   Exercises and solutions

**Exercise 10.1.** Does a convex function $f : \mathbb{R} \to \mathbb{R}$ with a unique global minimum have to be strictly convex? What if $f$ is differentiable?

**Solution 10.1.** A counterexample is the function $f : \mathbb{R} \to \mathbb{R}$ given by $f(x) = |x|$ which is convex but not strictly convex and has its unique global minimum at $x = 0$. If we have the differentiability requirement,

$$f(x) = \begin{cases} x^2 & \text{if } x \leq 1 \\ 2x - 1 & \text{if } x > 1 \end{cases}$$

is a counterexample.

★   ★   ★

**Exercise 10.2.** Let $f : \mathbb{R}^2 \to \mathbb{R}$ be a differentiable convex function and

$$S = \{(x_1, x_2) \in \mathbb{R}^2 \,|\, -1 \leq x_1 \leq 2, -1 \leq x_2 \leq 1\}.$$

Suppose that $\nabla f(x_0) = (1, 0)$ for $x_0 = (-1, \frac{1}{2})$. Prove that $x_0$ is a minimum for $f$ defined on $S$.

**Solution 10.2.** Since

$$\nabla f(x_0)(x - x_0) = (1, 0) \begin{pmatrix} x_1 + 1 \\ x_2 - \frac{1}{2} \end{pmatrix} = x_1 + 1 \geq 0 \qquad (10.3)$$

for every $x = (x_1, x_2) \in S$, the result follows from Proposition 10.1.

★   ★   ★

**Exercise 10.3.** Guess the solution to the optimization problem

$$\min\{(x-5)^2 + (y-5)^2 \mid x \geq 0,\ y \geq 0,\ x^2 + y^2 \leq 25\}. \qquad (10.4)$$

Show that your guess was correct!

**Solution 10.3.** Set $S = \{(x,y) \in \mathbb{R}^2 \mid x, y \geq 0,\ x^2 + y^2 \leq 25\}$. Geometrically, we want to find the point in $S$ that minimizes the distance to $(5,5)$. Thus, is seems like we should require that $x = y$ and $x^2 + y^2 = 25$ (see Figure 10.1). Specifically, we guess that the optimal point is

$$(x_0, y_0) = \left(\tfrac{5}{\sqrt{2}}, \tfrac{5}{\sqrt{2}}\right).$$

By Proposition 10.1 the guess is verified if

$$\left(\frac{10}{\sqrt{2}} - 10, \frac{10}{\sqrt{2}} - 10\right)\left(\begin{matrix} x - \frac{5}{\sqrt{2}} \\ y - \frac{5}{\sqrt{2}} \end{matrix}\right) = \left(\frac{10}{\sqrt{2}} - 10\right)\left(x + y - 5\sqrt{2}\right) \geq 0$$

for every $(x,y) \in S$. It suffices to show that $x + y \leq 5\sqrt{2}$ which is implied by convexity of $z \mapsto z^2$, since

$$(x+y)^2 \leq \tfrac{1}{2}(2x)^2 + \tfrac{1}{2}(2y)^2 = 2(x^2 + y^2) \leq 50,$$

for $(x,y) \in S$.

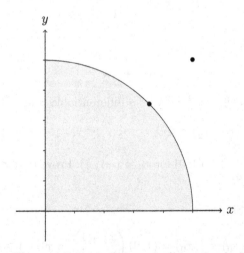

**Figure 10.1:** The set $S$ is shaded, and the optimal point $(x_0, y_0) = (5/\sqrt{2}, 5/\sqrt{2})$ and $(5,5)$ is marked.

★   ★   ★

**Exercise 10.4.** Let

$$S = \left\{ (x,y) \in \mathbb{R}^2 \;\middle|\; \begin{matrix} -x - y \leq 0 \\ 2x - y \leq 1 \\ -x + 2y \leq 1 \end{matrix} \right\}.$$

(i) Use the KKT conditions to solve the minimization problem

$$\min\{-x - 4y \,|\, (x,y) \in S\}.$$

(ii) Use the KKT conditions to solve the minimization problem

$$\min\{x + y \,|\, (x,y) \in S\}.$$

(iii) Solve the optimization problem

$$\max\{(x+1)^2 + (y+1)^2 \,|\, (x,y) \in S\}.$$

Give a geometric interpretation of your answer.

**Solution 10.4.** For convenience let $g_1(x,y) = -x - y$, $g_2(x,y) = 2x - y - 1$, and $g_3(x,y) = -x + 2y - 1$.

(i) After inspecting Figure 10.2 we use $(x_0, y_0) = (1,1)$ as a qualified guess for an optimal point. It follows by Theorem 10.6(4) that we verify that the guess is correct by checking that the KKT conditions hold at this point for some $\lambda_1, \lambda_2, \lambda_3 \in \mathbb{R}$. It must be that $\lambda_1 = 0$ if $g_1(1,1)\lambda_1 = 0$ and consequently,

$$\begin{pmatrix} -1 \\ -4 \end{pmatrix} + \lambda_2 \begin{pmatrix} 2 \\ -1 \end{pmatrix} + \lambda_3 \begin{pmatrix} -2 \\ 1 \end{pmatrix} = 0.$$

Solving the above system yields $(\lambda_2, \lambda_3) = (2,3)$ which has non-negative entries. In addition $(1,1) \in S$, showing that our guess is an optimal point.

(ii) Figure 10.2 gives the impression that the optimal points are exactly those on the form

$$\begin{pmatrix} x_\mu \\ y_\mu \end{pmatrix} = (1 - \mu) \begin{pmatrix} 1/3 \\ -1/3 \end{pmatrix} + \mu \begin{pmatrix} -1/3 \\ 1/3 \end{pmatrix} \in S \qquad (10.5)$$

for some $0 \leq \mu \leq 1$. Note that $g_1(x_\mu, y_\mu) = 0$. Since the system

$$\begin{pmatrix} 1 \\ 1 \end{pmatrix} + \lambda_1 \begin{pmatrix} -1 \\ -1 \end{pmatrix} = 0$$

has the solution $\lambda_1 = 1$, we conclude that KKT conditions are satisfied at $(x_\mu, y_\mu)$ with $\lambda_1 = 1$, $\lambda_2 = 0$, and $\lambda_3 = 0$. This shows that $(x_\mu, y_\mu)$ is optimal for $0 \le \mu \le 1$. Note that the optimal value is zero, and that this means $x = y$ in optimum. Consequently, we know that all optimal points are given as above.

(iii) Using the double description method in §5.3.1 (or just by computing the intersection of boundary lines) we write $S$ in its vertex representation

$$S = \text{conv}\left(\left\{\begin{pmatrix}1\\1\end{pmatrix}, \begin{pmatrix}-1/3\\1/3\end{pmatrix}, \begin{pmatrix}1/3\\-1/3\end{pmatrix}\right\}\right).$$

This representation is supported by Figure 10.2. Then Theorem 10.15 implies that the maximum is attained in one of the generating points of the convex hull. The largest function value is attained in $(1,1)$, hence this must be an optimal point. The optimal point is unique by the same theorem. Geometrically, $(1,1)$ can be interpreted as the point in $S$ having the greatest distance to $(-1,-1)$.

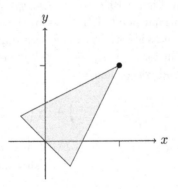

**Figure 10.2:** The set $S$ and the optimal point $(1,1)$ in (i) and (iii).

★ ★ ★

**Exercise 10.5.** Solve the optimization problem

$$\min\left\{x^2 + 2y^2 + 3z^2 - 2xz - xy \;\middle|\; \begin{matrix}2x^2 + y^2 + z^2 \le 4\\ x + y + z \le 1\end{matrix}\right\}.$$

**Solution 10.5.** Note that

$$x^2 + 2y^2 + 3z^2 - 2xz - xy = (x, y, z) \begin{pmatrix} 1 & -\frac{1}{2} & -1 \\ -\frac{1}{2} & 2 & 0 \\ -1 & 0 & 3 \end{pmatrix} \begin{pmatrix} x \\ y \\ z \end{pmatrix},$$

and since the associated matrix is positive definite by Theorem 9.11, the object function has a unique minimum on $\mathbb{R}^3$ in $(0, 0, 0)$. As this is also a feasible point, we have solved the problem.

**Exercise 10.6.** Let $S = \{(x, y) \mid 2x^2 + y^2 \leq 3, \ x^2 + 2y^2 \leq 3\}$ and $f(x, y) = (x - 4)^2 + (y - 4)^2$.

(i) State the KKT conditions for $\min\{f(x, y) \mid (x, y) \in S\}$ for $(x, y) = (1, 1)$.

(ii) Suppose now that $g(x, y) = (x - a)^2 + (y - b)^2$. For which $a$ and $b$ does $\min\{g(x, y) \mid (x, y) \in S\}$ have optimum in $(1, 1)$? State the KKT conditions when $(a, b) = (1, 1)$.

**Solution 10.6.** (i) In this setup we have that $g_1(x, y) = 2x^2 + y^2 - 3$ and $g_2(x, y) = x^2 + 2y^2 - 3$. For $(x, y) = (1, 1)$ the KKT conditions boil down to

$$\lambda_1, \lambda_2 \geq 0$$
$$\lambda_1(4, 2) + \lambda_2(2, 4) = (6, 6),$$

since $g_1(1, 1) = g_2(1, 1) = 0$.

(ii) The gradient condition is

$$\begin{pmatrix} 2 - 2a \\ 2 - 2b \end{pmatrix} + \lambda_1 \begin{pmatrix} 4 \\ 2 \end{pmatrix} + \lambda_2 \begin{pmatrix} 2 \\ 4 \end{pmatrix} = 0,$$

and the unique solution is given by

$$\begin{pmatrix} \lambda_1 \\ \lambda_2 \end{pmatrix} = \begin{pmatrix} \frac{2}{3}a - \frac{1}{3}b - \frac{1}{3} \\ -\frac{1}{3}a + \frac{2}{3}b - \frac{1}{3} \end{pmatrix}. \tag{10.6}$$

The KKT conditions at $(1, 1)$ are satisfied if and only if $\lambda_1, \lambda_2 \geq 0$ and (10.6) hold, and therefore

$$A = \{(a, b) \in \mathbb{R}^2 \mid -2a + b \leq -1, \ a - 2b \leq -1\}$$

is the set of $(a, b) \in \mathbb{R}^2$ with this property. If $(a, b) \in A$ it follows from Theorem 10.6(4) that $(1, 1)$ is an optimal solution, and if $(a, b) \notin A$ then Theorem 10.6(2) gives that $(1, 1)$ is not optimal. This shows that $A$ is the set of points for which $(1, 1)$ is optimal.

If $(a, b) = (1, 1)$ the KKT conditions read

$$\lambda_1, \lambda_2 \geq 0$$
$$2x^2 + y^2 - 3 \leq 0$$
$$x^2 + 2y^2 - 3 \leq 0$$
$$\lambda_1(2x^2 + y^2 - 3) = 0$$
$$\lambda_2(x^2 + 2y^2 - 3) = 0$$
$$2x - 2 + \lambda_1 4x + \lambda_2 2y = 0$$
$$2y - 2 + \lambda_1 2y + \lambda_2 4x = 0.$$

⋆ ⋆ ⋆

**Exercise 10.7.** Let $f : \mathbb{R}^2 \to \mathbb{R}$ be given by

$$f(x, y) = (x - 1)^2 + (y - 1)^2 + 2xy.$$

(i) Show that $f$ is a convex function.

(ii) Find $\min\{f(x, y) \mid (x, y) \in \mathbb{R}^2\}$. Is this minimum unique? Is $f$ a strictly convex function?

  Let

$$S = \{(x, y) \in \mathbb{R}^2 \mid x + y \leq 0,\ x - y \leq 0\}.$$

(iii) Apply the KKT-conditions to decide if $(-1, -1)$ is an optimal solution to

$$\min\{f(x, y) \mid (x, y) \in S\}.$$

(iv) Find

$$m = \min\{f(x, y) \mid (x, y) \in S\}$$

  and

$$\{(x, y) \in \mathbb{R}^2 \mid f(x, y) = m\}.$$

**Solution 10.7.**     (i) The Hessian

$$\nabla^2 f(x, y) = \begin{pmatrix} 2 & 2 \\ 2 & 2 \end{pmatrix},$$

is positive semidefinite for all $(x, y) \in \mathbb{R}^2$ by Example 9.16, and Theorem 9.5 implies that $f$ is convex.

(ii) Corollary 9.2 gives that $(x, y) \in \mathbb{R}^2$ is a global minimum if and only if

$$\nabla f(x, y) = (2(x + y) - 2, 2(x + y) - 2) = 0, \qquad (10.7)$$

which is satisfied if $x + y = 1$ and thus, the minimum is not unique. Lemma 7.6 implies that $f$ cannot be strictly convex.

(iii) Let $g_1(x, y) = x + y$ and $g_2(x, y) = x - y$. Since $g_1(-1, -1) = -2 < 0$, the KKT conditions imply $\lambda_1 = 0$. The system

$$\nabla f(-1, -1) + \lambda_2 \nabla g_2(-1, -1) = 0 \qquad (10.8)$$

has no solution, and therefore $(-1, -1)$ is not an optimal point by Theorem 10.6(3).

(iv) Consider a point $(x_0, y_0) \in \mathbb{R}^2$ such that $g_1(x_0, y_0) = 0$ and $g_2(x_0, y_0) < 0$. Then $y_0 = -x_0$ and $\lambda_2 = 0$. The equations

$$\nabla f(x_0, -x_0) + \lambda_1 \nabla g_1(x_0, -x_0) = 0,$$

are satisfied for $\lambda_1 = 2 \geq 0$, and therefore the KKT conditions hold at $(x_0, y_0)$ giving a solution to $\min\{f(x, y) \mid (x, y) \in S\}$. This shows that $m = f(x_0, y_0) = 2$. Finally,

$$\{(x, y) \in \mathbb{R}^2 \mid f(x, y) = 2\}$$
$$= \{(x, y) \in \mathbb{R}^2 \mid x + y = 2\} \cup \{(x, y) \in \mathbb{R}^2 \mid x + y = 0\}$$

which concludes the solution.

$$\star \quad \star \quad \star$$

**Exercise 10.8.** Let $f : \mathbb{R}^2 \to \mathbb{R}$ be given by

$$f(x, y) = x^2 + y^2 - e^{x - y - 1}$$

and let

$$C = \{(x, y) \mid x - y \leq 0\}.$$

(i) Show that $f : \mathbb{R}^2 \to \mathbb{R}$ is not a convex function.

(ii) Show that $f$ is a convex function on the open subset

$$\{(x, y) \in \mathbb{R}^2 \mid x - y < \tfrac{1}{2}\}$$

and conclude that $f$ is convex on $C$.

(iii) Show that $v = (0,0)$ is an optimal solution for the optimization problem $\min\{f(v) \mid v \in C\}$. Is $v$ a unique optimal solution here?

**Solution 10.8.**     (i) The Hessian of $f$ is

$$\nabla^2 f(x,y) = \begin{pmatrix} 2 - e^{x-y-1} & e^{x-y-1} \\ e^{x-y-1} & 2 - e^{x-y-1} \end{pmatrix}.$$

Observe that the first leading principal minor is negative for $x \gg y$. From this we use Theorem 9.5 and Theorem 9.15 to conclude that $f$ is not a convex function on $\mathbb{R}^2$.

(ii) Note that $e^{x-y-1} < e^0 = 1$ when $x - y < 1/2$. Therefore $\nabla^2 f(x,y)$ is positive definite on

$$U = \left\{ (x,y) \in \mathbb{R}^2 \mid x - y < \frac{1}{2} \right\},$$

and since $U$ is open and convex, $f$ is strictly convex on $U$ by Theorem 9.5. The inclusion $C \subseteq U$ implies that $f$ is strictly convex on $C$ as well.

(iii) From Proposition 10.1 we get that $v = (0,0)$ is an optimal solution, since

$$\nabla f(v)\left( \begin{pmatrix} x \\ y \end{pmatrix} - v \right) = e^{-1}(y - x) \geq 0$$

for $(x,y) \in C$. We deduce from Lemma 7.6 that this optimum is unique since $f$ is strictly convex on $C$.

$$\star \quad \star \quad \star$$

**Exercise 10.9.** Let $f : \mathbb{R}^4 \to \mathbb{R}$ be given by

$$f(x_1, x_2, x_3, x_4) = (x_1 - x_3)^2 + (x_2 - x_4)^2$$

and $C \subseteq \mathbb{R}^4$ by

$$C = \{(x_1, x_2, x_3, x_4) \in \mathbb{R}^4 \mid x_1^2 + (x_2 - 2)^2 \leq 1,\ x_3 - x_4 \geq 0\}.$$

(i) Show that $f$ is a convex function. Is $f$ strictly convex?
(ii) Show that $C$ is a convex subset of $\mathbb{R}^4$.
(iii) Does there exist an optimal point $v = (x_1, x_2, x_3, x_4) \in \mathbb{R}^4$ for the minimization problem

$$\min_{v \in C} f(v)$$

with $x_3 = x_4 = 0$?

(iv) Does there exist an optimal point $v = (x_1, x_2, x_3, x_4) \in \mathbb{R}^4$ for the minimization problem

$$\min_{v \in C} f(v)$$

with $x_3 = x_4 = 1$?

**Solution 10.9.** Before solving the exercise we give a geometric interpretation of the optimization problem $\min_{v \in C} f(v)$. First notice that $f(x_1, x_2, x_3, x_4)$ gives the (squared) distance between $(x_1, x_2)$ and $(x_3, x_4)$. Imposing the restriction that points should belong to $C$ is equivalent to have $(x_1, x_2)$ in the unit disc centered at $(0, 2)$ and $(x_3, x_4)$ below the positive (sloping) diagonal of $\mathbb{R}^2$. As a consequence, an optimal solution consists of a pair of points in $\mathbb{R}^2$ having a minimal distance to each other. Figure 10.3 depicts a feasible (however not optimal) pair of such points.

(i) First, we compute that

$$\nabla^2 f(x) = \begin{pmatrix} 2 & 0 & -2 & 0 \\ 0 & 2 & 0 & -2 \\ -2 & 0 & 2 & 0 \\ 0 & -2 & 0 & 2 \end{pmatrix}$$

for $x \in \mathbb{R}^4$. Theorem 9.5 tells us that $f$ is convex if and only if $\nabla^2 f(x)$ is positive semidefinite. We can make use of the procedure given in the proof of Theorem 9.17 (and in Example 9.19) and find

$$B_2^t \left( B_1^t \left( \nabla^2 f(x) \right) B_1 \right) B_2 = \begin{pmatrix} 2 & 0 & 0 & 0 \\ 0 & 2 & 0 & 0 \\ 0 & 0 & 0 & 0 \\ 0 & 0 & 0 & 0 \end{pmatrix},$$

where

$$B_1 = \begin{pmatrix} 1 & 0 & 1 & 0 \\ 0 & 1 & 0 & 0 \\ 0 & 0 & 1 & 0 \\ 0 & 0 & 0 & 1 \end{pmatrix} \quad \text{and} \quad B_2 = \begin{pmatrix} 1 & 0 & 0 & 0 \\ 0 & 1 & 0 & 1 \\ 0 & 0 & 1 & 0 \\ 0 & 0 & 0 & 1 \end{pmatrix}.$$

Now it follows from Lemma 9.6 that $\nabla^2 f(x)$ is positive semidefinite. We find that $f$ is not strictly convex since the function is constant along points where $x_1 = x_3$ and $x_2 = x_4$.

(ii) Since $C$ is an intersection of a set of the form $\{x \in \mathbb{R}^4 \mid g(x) \leq 0\}$ where $g : \mathbb{R}^4 \to \mathbb{R}$ is a convex function and a polyhedron, $C$ is a convex subset.

(iii) For a given point $(x_1, x_2, 0, 0) \in \mathbb{R}^4$ the gradient KKT condition reads

$$\begin{pmatrix} 2x_1 \\ 2x_2 \\ -2x_1 \\ -2x_2 \end{pmatrix} + \lambda_1 \begin{pmatrix} 2x_1 \\ 2x_2 - 4 \\ 0 \\ 0 \end{pmatrix} + \lambda_2 \begin{pmatrix} 0 \\ 0 \\ -1 \\ 1 \end{pmatrix} = 0. \qquad (10.9)$$

If the KKT conditions should be satisfied, $\lambda_1, \lambda_2 \geq 0$, so the first entrance in (10.9) implies that $x_1 = 0$. The third and fourth entrance gives that $\lambda_2 = 0$ and $x_2 = 0$. As $(0, 0, 0, 0) \notin C$, no point with $x_3 = x_4 = 0$ satisfies the KKT conditions, hence nu such point is optimal by Theorem 10.6(2).

(iv) If the point should be optimal when fixing $x_3 = x_4 = 1$ we need to minimize the function $(z_1, z_2) = (z_1 - 1)^2 + (z_2 - 1)^2$ on the set

$$\left\{ (x_1, x_2) \in \mathbb{R}^2 \mid x_1^2 + (x_2 - 2)^2 \leq 1 \right\}.$$

Due to our initial discussion, a natural guess of an optimal point will be

$$\begin{pmatrix} x_1 \\ x_2 \end{pmatrix} = \begin{pmatrix} \cos(-\pi/4) \\ 2 + \sin(-\pi/4) \end{pmatrix} = \begin{pmatrix} 1/\sqrt{2} \\ 2 - 1/\sqrt{2} \end{pmatrix}$$

and thus, we want to check if $z = (1/\sqrt{2}, 2 - 1/\sqrt{2}, 1, 1)$ satisfies the KKT conditions. After checking that $z \in C$, one verifies that the gradient condition for $z$ reads

$$\begin{pmatrix} \sqrt{2} - 2 \\ 2 - \sqrt{2} \\ 2 - \sqrt{2} \\ \sqrt{2} - 2 \end{pmatrix} + \lambda_1 \begin{pmatrix} \sqrt{2} \\ -\sqrt{2} \\ 0 \\ 0 \end{pmatrix} + \lambda_2 \begin{pmatrix} 0 \\ 0 \\ -1 \\ 1 \end{pmatrix} = 0.$$

As we see that $\lambda_1, \lambda_2 \geq 0$, the KKT conditions are all satisfied, and since requirements for applying Theorem 10.6(4) holds, $z$ is an optimal solution for the minimization problem.

$$\star \quad \star \quad \star$$

**Exercise 10.10.** Let

$$f(x, y) = (x - 1)^2 + y^2$$

and

$$C = \{ (x, y) \in \mathbb{R}^2 \mid -1 \leq x \leq 0, \ -1 \leq y \leq 1 \}.$$

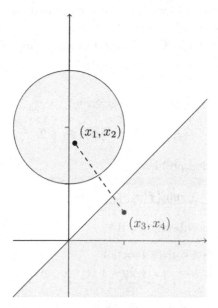

**Figure 10.3:** A geometric interpretation of Exercise 10.9.

(i) Solve the optimization problem

$$\min\{f(v) \,|\, v \in C\}.$$

(ii) Solve the optimization problem

$$\max\{f(v) \,|\, v \in C\}.$$

How many optimal solutions are there? Why?

**Solution 10.10.** (i) Note that $f$ is strictly convex. This means that we only have to search for one solution to the minimization problem as this would be unique (see Lemma 7.6). Using Proposition 10.1 it is quickly verified that $(0,0)$ is the optimal solution, since

$$\nabla f(0,0) \begin{pmatrix} x \\ y \end{pmatrix} = (-2,0) \begin{pmatrix} x \\ y \end{pmatrix} = -2x \geq 0$$

whenever $(x,y) \in C$.

(ii) We note that

$$C = \text{conv}\left(\left\{\begin{pmatrix} -1 \\ -1 \end{pmatrix}, \begin{pmatrix} -1 \\ 1 \end{pmatrix}, \begin{pmatrix} 0 \\ -1 \end{pmatrix}, \begin{pmatrix} 0 \\ 1 \end{pmatrix}\right\}\right).$$

From Theorem 10.15 we find that the optimal solutions to the problem $\max\{f(v)|v \in C\}$ are $(-1, -1)$ and $(-1, 1)$. These are the only optimal solutions by Corollary 7.4 (Jensen's inequality).

$$\star \quad \star \quad \star$$

**Exercise 10.11.** Let $f : \mathbb{R}^2 \to \mathbb{R}$ be given by

$$f(x, y) = \tfrac{1}{2}x^2 + y^2 - 2y + 2.$$

Below, the minimization problem

$$\min\{f(x, y) \,|\, (x, y) \in S\} \tag{10.10}$$

is analyzed for various subsets $S \subseteq \mathbb{R}^2$.

(i) Show that $f$ is a convex function
(ii) Show that $h(x, y) = -x + 2y - 1$ is a convex and concave function.
(iii) Let
$$S = \{(x, y) \in \mathbb{R}^2 \,|\, -x + 2y \leq 1\}.$$

Show that $(-1, 0) \in S$ cannot be an optimal solution to (10.10). Find an optimal solution to (10.10).
(iv) Find an optimal solution in (10.10) for

$$S = \{(x, y) \in \mathbb{R}^2 \,|\, -x + 2y \geq 1\}.$$

(v) Are the optimal solutions in ((iii)) and ((iv)) unique?

**Solution 10.11.**      (i) The Hessian matrix of $f$ is given by

$$\nabla^2 f(x, y) = \begin{pmatrix} 1 & 0 \\ 0 & 2 \end{pmatrix}.$$

This is a positive definite matrix by Lemma 9.6, and Theorem 9.5 implies that $f$ is strictly convex.
(ii) We find that

$$\nabla^2 h = \nabla^2(-h) = \begin{pmatrix} 0 & 0 \\ 0 & 0 \end{pmatrix},$$

and another use of Theorem 9.5 yields that both $h$ and $-h$ are convex.

(iii) Since $(1,1) \in S$ and

$$\nabla f(-1,0)\left(\begin{pmatrix} 1 \\ 1 \end{pmatrix} - \begin{pmatrix} -1 \\ 0 \end{pmatrix}\right) = (-1,-2)\begin{pmatrix} 2 \\ 1 \end{pmatrix} = -4,$$

Proposition 10.1 implies that $(-1,0)$ is not optimal.

To find an optimal solution to $\min\{f(x,y) \mid (x,y) \in S\}$ we find a point that satisfies the KKT conditions. First we assume that the restriction defining $S$ is binding, $-x+2y = 1$. If this is the case, the gradient condition reads

$$\begin{pmatrix} x \\ 2y-2 \end{pmatrix} + \lambda \begin{pmatrix} -1 \\ 2 \end{pmatrix} = 0.$$

The solution to these equations is $(x,y,\lambda) = (1/3, 2/3, 1/3)$, in particular the third entrance is non-negative, which means that $(1/3, 2/3)$ is an optimal point for $\min\{f(x,y) \mid (x,y) \in S\}$.

(iv) Solving $\nabla f(x,y) = 0$ gives $(x,y) = (0,1) \in S$, thus we have an optimal solution to $\min\{f(x,y) \mid (x,y) \in S\}$, since $f$ is convex by (i).

(v) Both solutions are unique by Lemma 7.6.

$$\star \quad \star \quad \star$$

**Exercise 10.12.** Let $T$ denote the convex hull of $(1,1), (-1,2), (2,3) \in \mathbb{R}^2$.

(i) Solve the optimization problem

$$\max\{x^2 + y^3 \mid (x,y) \in T\}.$$

Is your solution unique?

(ii) Solve the optimization problem

$$\min\{x^2 + y^2 \mid (x,y) \in T\} \tag{10.11}$$

and give a geometric interpretation.

**Solution 10.12.**     (i) The Hessian matrix of $(x,y) \mapsto x^2 + y^3$ is

$$\begin{pmatrix} 2 & 0 \\ 0 & 6y \end{pmatrix}$$

and thus, the function is strictly convex on $T$ by Theorem 9.5 since $y \geq 1$ for every $(x,y) \in T$. Evaluating the function in $(1,1)$, $(-1,2)$, and $(2,3)$ gives function values of 2, 9, and 31 respectively, and Theorem 10.15 implies that the unique optimal point is $(2,3)$.

(ii) By considering Figure 10.4 (or using the double description method described in §5.3) we get the representation

$$T = \left\{ (x,y) \in \mathbb{R}^2 \mid \tfrac{3}{2} - \tfrac{1}{2}x \leq y, \tfrac{5}{2} + \tfrac{1}{2}x \geq y, -1 + 2x \leq y \right\}.$$

In order to search for a point satisfying the KKT conditions we consider Figure 10.4 and guess that such a point has $3/2 - x/2 = y$ (since we are trying to find the shortest distance from $(0,0)$ to $T$). Assuming this holds, the gradient condition reads

$$(2x, 2y) + \lambda \left( -\tfrac{1}{2}, 1 \right) = 0.$$

The equations have the solution $(x_0, y_0, \lambda) = (3/5, 6/5, 12/5)$ where $(x_0, y_0) \in T$, and we conclude that this is a (unique) minimum by Theorem 10.6(4).

For a geometric interpretation let $d \geq 0$ and note that points $(x, y) \in \mathbb{R}^2$ satisfying

$$x^2 + y^2 = d$$

represent those that have a distance of $\sqrt{d}$ to $(0,0)$. As a consequence, looking for a point from a given set that minimizes $d$ can be thought of as finding the point in the set with shortest (Euclidean) distance to $(0,0)$.

$$\star \quad \star \quad \star$$

**Exercise 10.13.** Let $C$ denote the set of points $(x, y) \in \mathbb{R}^2$ such that

$$x^2 + 2y^2 \leq 1$$
$$x + y \geq 1 \qquad\qquad (10.12)$$
$$y \leq x.$$

(i) Show that $C$ is a convex set and that there exists $(x_0, y_0) \in \mathbb{R}^2$ such that

$$x_0^2 + 2y_0^2 < 1$$
$$x_0 + y_0 > 1$$
$$y_0 < x_0.$$

(ii) Solve the optimization problem

$$\max\{x + 3y \mid (x, y) \in C\}.$$

Is your solution unique?

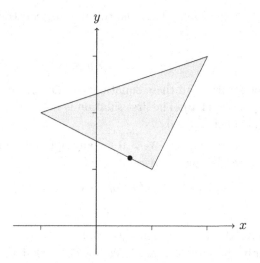

**Figure 10.4:** The convex hull $T$ and the optimal solution $(x_0, y_0) = (3/5, 6/5)$ marked.

**Solution 10.13.**    (i) The set $C$ is convex as the intersection of a polyhedron and a set of the form $\{(x,y) \in \mathbb{R}^2 \mid g(x,y) \leq 0\}$ for a convex function $g$. A strictly feasible point could be $(x_0, y_0) = (4/5, 1/4)$.

(ii) Initially, note that the task corresponds to finding optimal solutions for $\min\{-x-3y \mid (x,y) \in C\}$. By Theorem 10.6 the set of solutions to the optimization problem is characterized by the points satisfying the associated KKT conditions (note that the problem is strictly feasible by (i)).

The gradient condition reads

$$-1 + 2\lambda_1 x - \lambda_2 - \lambda_3 = 0 \tag{10.13}$$
$$-3 + 4\lambda_1 y - \lambda_2 + \lambda_3 = 0, \tag{10.14}$$

from which we may conclude that $\lambda_1 > 0$ by (10.13). In fact, it also holds that $\lambda_3 > 0$. To see this assume that $\lambda_3 = 0$ and consider the two cases:

(a) If $\lambda_2 = 0$ then (10.13) and (10.14) boil down to

$$\lambda_1 x = \tfrac{1}{2}$$
$$\lambda_1 y = \tfrac{3}{4},$$

and since $\lambda_1 > 0$ this implies that $y > x$ which no element in $C$ satisfies.

(b) If $\lambda_2 > 0$ we must have the two binding constraints

$$x^2 + 2y^2 = 1$$
$$x + y = 1.$$

The solutions to these equations are $(x_1, y_1) = (1/3, 2/3)$ and $(x_2, y_2) = (1, 0)$. The first solution is not in $C$ and the second contradicts $\lambda_2 > 0$.

Summing up we have $\lambda_1, \lambda_3 > 0$ for any optimal solution, and the corresponding binding constraints read

$$x^2 + 2y^2 = 1$$
$$x = y,$$

which have the two solutions $(1/\sqrt{3}, 1/\sqrt{3})$ and $(-1/\sqrt{3}, -1/\sqrt{3})$ where only the former is in $C$. We conclude that this point is the unique solution.

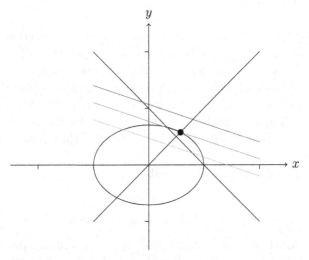

**Figure 10.5:** The convex set $C$ defined in (10.12), different contours for the object function $-f$ and the optimal solution $(x_0, y_0) = (1/\sqrt{3}, 1/\sqrt{3})$ marked.

⋆  ⋆  ⋆

**Exercise 10.14.** Let $f : \mathbb{R}^2 \to \mathbb{R}$ be given by

$$f(x, y) = x^2 + y^4 + xy.$$

(i) Is $f$ a convex function?

(ii) Let $F$ denote the convex hull of $(-1,1)$, $(-1,2)$, $(-2,2)$, $(-3,1) \in \mathbb{R}^2$. Solve the optimization problem

$$\max\{f(x,y) \mid (x,y) \in F\}.$$

Is your maximum unique?

(iii) Show in detail that

$$\min\{f(x,y) \mid (x,y) \in F\} \tag{10.15}$$

has the unique solution $x = -1$ og $y = 1$.

(iv) Does $f$ have a unique global minimum on $\mathbb{R}^2$?

**Solution 10.14.** (i) Since

$$\nabla^2 f(x,y) = \begin{pmatrix} 2 & 1 \\ 1 & 12y^2 \end{pmatrix}, \tag{10.16}$$

is not positive semidefinite whenever $y \in (-\frac{1}{2\sqrt{6}}, \frac{1}{2\sqrt{6}})$, $f$ is not convex on $\mathbb{R}^2$ (by Theorem 9.5).

(ii) Since $y \geq 1$ for all $(x,y) \in F$ then $f$ is convex on $F$ by (i), and the calculations

$$f(-1,1) = 1, \quad f(-1,2) = 15, \quad f(-2,2) = 16, \quad \text{and} \quad f(-3,1) = 7$$

gives that $\max\{f(x,y) \mid (x,y) \in F\} = 16$ and that the unique solution is $(x_0, y_0) = (-2,2)$ cf. Theorem 10.15.

(iii) From Figure 10.6 (or using the double description method from §5.3) we may deduce that

$$F = \{(x,y) \in \mathbb{R}^2 \mid x + 1 \leq 0,\ 1 \leq y \leq 2,\ y - x - 4 \leq 0\}.$$

We want to show that $(-1,1)$ satisfies the KKT conditions since this would imply optimality by Theorem 10.6(4). First, note that $(-1,1) \in F$. In this point two of the four constraint in $F$ will be binding, and therefore the gradient condition is

$$\begin{pmatrix} -1 \\ 3 \end{pmatrix} + \lambda \begin{pmatrix} 1 \\ 0 \end{pmatrix} + \mu \begin{pmatrix} 0 \\ -1 \end{pmatrix} = 0.$$

The unique solution to this system is $\lambda = 1$ and $\mu = 3$ and since these numbers are non-negative, we conclude that all of the KKT conditions are satisfied in $(-1,1)$.

It follows from (10.16) that $f$ is strictly convex on $F$, so by Lemma 7.6 we conclude that $(-1,1)$ is the unique solution to (10.15).

(iv) First, note that $f(0,0) = 0$ and $f(-1/2, 3/4) = -5/256$, so $(0,0)$ cannot be a global minimum on $\mathbb{R}^2$. Second, we see that $f(x,y) = f(-x, -y)$, and therefore if any point is a global minimum, then the point with opposite sign would be as well. We conclude that there cannot be a unique global minimum.

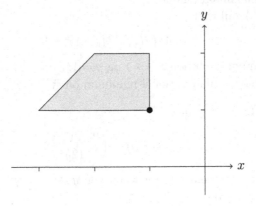

**Figure 10.6:** The convex hull $F$ in grey and the optimal solution $(x_0, y_0) = (-1, 1)$ marked.

$$\star \quad \star \quad \star$$

**Exercise 10.15.** Let $f : \mathbb{R}^2 \to \mathbb{R}$ be given by

$$f(x,y) = x^2 + xy + y^2$$

and let

$$S = \left\{ (x,y) \in \mathbb{R}^2 \,\middle|\, \begin{array}{c} y - x \geq 1 \\ y + x \geq 1 \\ y \leq 2 \end{array} \right\}.$$

(i) Is $(0,1)$ an optimal solution to

$$\min\{f(x,y) \,|\, (x,y) \in S\}?$$

(ii) Find

$$\max\{f(x,y) \,|\, (x,y) \in S\}.$$

**Solution 10.15.**    (i) We will verify that $(0,1)$ satisfies the KKT conditions; then Theorem 10.6(4) will imply that it is the minimum. In the point $(0, 1) \in S$, the two first constraints bind and consequently, the gradient condition reads

$$\begin{pmatrix} 1 \\ 2 \end{pmatrix} + \lambda \begin{pmatrix} 1 \\ -1 \end{pmatrix} + \mu \begin{pmatrix} -1 \\ -1 \end{pmatrix} = 0.$$

The solution to the above system is $\lambda = 1/2$ and $\mu = 3/2$, and we see that the KKT conditions are satisfied. Furthermore, this minimum is unique by strict convexity of $f$.

(ii) In pursuit of using Theorem 10.15, we rewrite $S$ as

$$S = \mathrm{conv}(\{(0, 1), (-1, 2), (1, 2)\})$$

(see Figure 10.7 or use the double description method from §5.3). Since

$$f(0, 1) = 1, \quad f(-1, 2) = 3, \quad \text{and} \quad f(1, 2) = 7,$$

$f$ attains its unique maximum on $S$ at $(1, 2)$ with a value of 7.

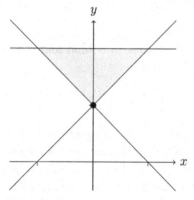

**Figure 10.7:** The polyhedron $S$ from Exercise 10.15 in grey and the optimal solution $(0, 1)$ marked.

⋆ ⋆ ⋆

**Exercise 10.16.** Let $f : \mathbb{R}^2 \to \mathbb{R}$ be given by

$$f(x, y) = 2x^2 + 3x + y^2 + y.$$

(i) Show that $f$ is a convex function and solve the minimization problem
   $\min\{f(x,y)\,|\,x,y\in\mathbb{R}\}$.
   Now let

$$S = \left\{(x,y)\in\mathbb{R}^2 \,\middle|\, \begin{matrix} x^2+(y+1)^2 \\ y \end{matrix} \begin{matrix} \leq 1 \\ \leq 0 \end{matrix} \, x \right\}$$

and consider the minimization problem (P) given by

$$\min\{f(x,y)\,|\,(x,y)\in S\}.$$

(ii) Show using the KKT conditions that $(0,0)$ is not optimal for (P).
(iii) Find an optimal solution for (P). Is it unique?

**Solution 10.16.**     (i) By Theorem 9.5 and Lemma 9.6, $f$ is strictly convex since

$$\nabla^2 f(x,y) = \begin{pmatrix} 4 & 0 \\ 0 & 2 \end{pmatrix}.$$

To solve $\min\{f(x,y)\,|\,x,y\in\mathbb{R}\}$ recall that Corollary 9.2 ensures that we just need to find $(x,y)$ such that $\nabla f(x,y) = 0$. We calculate

$$\nabla f(x,y) = (4x+3, 2y+1) = 0$$

which has the solution $(x,y) = (-3/4, -1/2)$.

(ii) At $(0,0)$ both constraints in $S$ bind and consequently, the (KKT) gradient condition reads

$$\begin{pmatrix} 3 \\ 1 \end{pmatrix} + \lambda \begin{pmatrix} 0 \\ 2 \end{pmatrix} + \mu \begin{pmatrix} -1 \\ 1 \end{pmatrix} = 0.$$

The solution to the system of equations is $\lambda = -2$ and $\mu = 3$, and we see that the KKT conditions are not satisfied.

(iii) To get a first grasp of finding an optimal solution we start by considering (i). Here we found the optimal point $(-3/4, -1/2)$. As a purely heuristic exercise, we could note that this point only violates the second constraint in $S$. Therefore, considering a point with $x = y$ may be fruitful. Building on this, we see that $f(x,x) = 3x^2 + 4x$ attains its minimum value in $x = -2/3$ which motivates the guess $(-2/3, -2/3)$ as an optimal point. Note that the first constraint is not binding and therefore the (KKT) gradient condition is

$$\begin{pmatrix} 1/3 \\ -1/3 \end{pmatrix} + \lambda \begin{pmatrix} -1 \\ 1 \end{pmatrix} = 0.$$

This is satisfied for $\lambda = 1/3$ and we conclude that $(-2/3, -2/3)$ is an optimal point for (P). Since $f$ is strictly convex (as noted in (i)), the optimum is unique.

$$\star \quad \star \quad \star$$

**Exercise 10.17.** Let $f : \mathbb{R}^2 \to \mathbb{R}$ be a differentiable function and suppose that $(x_0, y_0)$ is a saddle point of $f$ in the sense that

$$f(x_0, y_0) \le f(x, y_0) \text{ for every } x \in \mathbb{R}$$

$$f(x_0, y) \le f(x_0, y_0) \qquad \text{for every } y \in \mathbb{R}.$$

Prove that $(x_0, y_0)$ is a critical point for $f$. Show that a local extremum for $f$ is not a saddle point. Is a critical point for $f$, which is not an extremum a saddle point in the above sense?

**Solution 10.17.** Assume that $(x_0.y_0)$ is a saddle point of $f$. Consider the differentiable function $x \mapsto f(x, y_0)$, and note that it has a global minimum at $x_0$. We conclude by Lemma 7.11 that it is a critical point. Similarly, $y_0$ is a critical point for $y \mapsto f(x_0, y)$. Consequently,

$$\nabla f(x_0, y_0) = \left( \frac{d}{dx} f(x_0, y_0), \frac{d}{dy} f(x_0, y_0) \right) = 0.$$

For a constant function every point is a local extremum and a saddle point. In the case where $(x_0, y_0)$ is a strict local minimum[1], the inequality $f(x_0, y) \le f(x_0, y_0)$ fails for some $y$ which contradicts that $(x_0, y_0)$ is a saddle point. Similarly if $(x_0, y_0)$ is a strict local maximum. Consequently, a point cannot both be a saddle point and a strict local extremum.

The function $f(x, y) = x^3/3 - x - y^2$ has $\nabla f(1, 0) = 0$, but $(1, 0)$ is neither a saddle point (in the above sense) nor an extremum.

$$\star \quad \star \quad \star$$

**Exercise 10.18.** Pick two optimization problems from the exercises in this section and compute their dual problems. Solve the dual problems and compare with the solution of the original problems.

---

[1] We interpret a strict local minimum as a point $(x_0, y_0)$ for which there exists $\epsilon > 0$ such that $f(x_0, y_0) < f(x, y)$ for every $(x, y) \ne (x_0, y_0)$ with $|(x_0, y_0) - (x, y)| < \epsilon$.

**Solution 10.18.** We will solve the dual to the problem introduced in Exercise 10.7. In this case the Lagrangian reads

$$\mathcal{L}(x, y, \lambda_1, \lambda_2) = (x-1)^2 + (y-1)^2 + 2xy + \lambda_1(x+y) + \lambda_2(x-y)$$
$$= (x+y-1)^2 + 1 + \lambda_1(x+y) + \lambda_2(x-y).$$

We see that $q(\lambda) = -\infty$ if $\lambda_2 > 0$ and consequently, an optimal solution can be found with $\lambda_2 = 0$. The convex function $(x, y) \mapsto \mathcal{L}(x, y, \lambda_1, 0)$ attains its minimum when $x + y = 1 - \lambda_1/2$. We conclude that

$$q(\lambda_1, 0) = -\tfrac{1}{4}\lambda_1^2 + \lambda_1 + 1,$$

which is maximized in $\lambda_1^* = 2$ with a resulting value of $q^* = 2$.

The second problem we consider is the one introduced in Exercise 10.11 with $S = \{(x, y) \in \mathbb{R}^2 \mid -x + 2y \leq 1\}$, in which case the Lagrangian is given by

$$\mathcal{L}(x, y, \lambda) = \tfrac{1}{2}x^2 + y^2 - 2y + 2 + \lambda(2y - x - 1).$$

For a fixed $\lambda \geq 0$ we find that the convex function $(x, y) \mapsto \mathcal{L}(x, y, \lambda)$ attains its global minimum when $x = \lambda$ and $y = 1 - \lambda$. Consequently,

$$q(\lambda) = \mathcal{L}(\lambda, 1 - \lambda, \lambda) = -\tfrac{3}{2}\lambda^2 + \lambda + 1,$$

and we find that the solution to the dual problem $\sup\{q(\lambda) \mid \lambda \geq 0\}$ is $\lambda^* = 1/3$ with $q^* = q(\lambda^*) = 7/6$.

We note that the optimal values of the dual problems coincide with the optimal values of the original problems (as predicted by Theorem 10.9), and that the optimal points are the associated Lagrange multipliers.

$$\star \quad \star \quad \star$$

**Exercise 10.19.** Go through the steps of the interior point algorithm in §10.5 for the basic example $\min\{x \mid 0 \leq x \leq 1\}$.

**Solution 10.19.** In line with notation from §10.5 we define, for $\epsilon > 0$,

$$f_\epsilon(x) = x - \epsilon(\log(x) + \log(1 - x))$$

when $x \in (0, 1)$ and consequently,

$$f_\epsilon'(x) = 1 - \epsilon\left(\frac{1}{x} + \frac{1}{x - 1}\right).$$

The equation $\nabla f_\epsilon(x) = 0$ is solved for

$$x_\epsilon = \frac{1 + 2\epsilon - \sqrt{1 + 4\epsilon^2}}{2}.$$

Since $x_\epsilon \to 0$ when $\epsilon \to 0$, Theorem 10.11 gives $\min\{x \mid 0 \le x \le 1\} = 0$.

$\star$   $\star$   $\star$

**Exercise 10.20.** Consider the setup in §10.5. Prove that $f_\epsilon$ is a convex function on $S^o$ for $\epsilon > 0$. Is the Hessian of $f_\epsilon$ always positive definite?

**Solution 10.20.** Since $x \mapsto -\log(-x)$ is convex and increasing on $(-\infty, 0)$ and $g_i$ is convex, (a straightforward generalization of) Exercise 7.19 implies that $x \mapsto -\log(-g_i(x))$ is convex. Exercise 7.9 gives that the set of convex functions is a convex cone which implies that $f_\epsilon$ is convex.

An example of an optimization problem where the Hessian of $f_\epsilon$ is not positive definite is

$$\min\{x \mid x^4 \le 1\}.$$

We find that $f_\epsilon$ is defined on $(-1, 1)$ and

$$f_\epsilon''(x) = \epsilon \left( \frac{12x^2}{x^4 - 1} - \frac{16x^6}{(1 - x^4)^2} \right),$$

which is not strictly positive for all $x \in (-1, 1)$.

$\star$   $\star$   $\star$

**Exercise 10.21.** Implement the interior point algorithm in §10.5 in your favorite language (C, Haskell, *Mathematica*, ...). Test your implementation on the examples in §10.5.

**Solution 10.21.** The interior point algorithm is implemented in MATLAB (version R2015b). The examples we are considering involve only polyhedral constraints and as a consequence, we can use the formulas from §10.5.2. The function fEpsOpt implemented below returns the approximative solution to $\nabla f_\epsilon(x) = 0$ as it goes through the steps in §10.5.1 for a given (positive) limit on $\delta$.

```
%Function that returns the approximative minimum for f_eps.
function [x_eps]=fEpsOpt(A,b,x0,grad,hess,eps,delta_lim)
```

```
%Computing the gradient and hessian of f_epsilon.
grad_fEps=@(x) grad(x)+eps*((b-A*(x')).^(-1))'*A;
hess_fEps=@(x) hess(x)+eps*A'*diag((b-A*(x')).^(-2))*A;

%Newton-Raphson adjustment.
v=-hess_fEps(x0')^(-1)*grad_fEps(x0')';

%Computing [U.C.,(10.21)] for polyhedral constraints.
C=(b-A*x0)./(A*v); t0=min(C(A*v>0));

g=@(t) grad_fEps(x0'+t.*v')*v;

%Initial step.
delta=t0/2; t1=0; t2=0;
while g(t2)<0
    t1=t2; t2=t2+delta; delta=delta/2;
end

%Classical bisection algorithm.
while delta>delta_lim
    t=t1+delta;
    if g(t)<0
        t1=t;
    else
        t2=t;
    end
    delta=delta/2;
end
t_mid=t1+delta/2;
x_eps=x0+t_mid*v;
end
```

Having this function, we use the iterative procedure in [U.C., (10.22)] to compute $x_{\epsilon_N}$ for a large $N$ and some $0 < k < 1$. We call this function intPoint_method and it will return both $x_{\epsilon_N}$ and $f(x_{\epsilon_N})$. Due to Theorem 10.18, $f(x_{\epsilon_N})$ is approximately the value of [U.C., (10.18)] and provided that there exists a unique optimal point, $x_{\epsilon_N}$ will approximate that one.

```
function [x_eps,f_opt]=...
    intPoint_method(A,b,x0,f,grad,hess,eps,delta_lim,N,k)

%Approximate solution to [U.C.,(10.19)] for the initial choice of epsilon.
x_eps=fEpsOpt(A,b,x0,grad,hess,eps,delta_lim);

%Iterating in epsilon.
N=50;
k=0.5;
x_eps=fEpsOpt(A,b,x0,grad,hess,eps,delta_lim);
```

```
for j=1:N
    eps=k*eps;
    x_eps=fEpsOpt(A,b,x_eps,grad,hess,eps,delta_lim);
end

%Function value for the N-th iteration in epsilon.
f_opt=f(x_eps);
end
```

# Appendix A

# Analysis

## A.1 Introduction

The exercises in this appendix are meant to help the reader become more familiar with concepts from analysis which are essential in the main material. The focus is on convergence of sequences and the interior, boundary, and closure of sets, together with some useful inequalities.

**Theorem A.2.** *For two vectors $x, y \in \mathbb{R}^n$ the inequality*

$$|x + y| \leq |x| + |y|$$

*holds.*

**Definition A.14.** A subset $F \subseteq \mathbb{R}^n$ is called closed if every convergent sequence $(x_n) \subseteq F$ converging to $x \in \mathbb{R}^n$ has $x \in F$. A subset $U \subseteq \mathbb{R}^n$ is called open if its complement $\mathbb{R}^n \setminus U$ is closed.

**Definition A.15.** The closure of a subset $S \subseteq \mathbb{R}^n$ is defined as

$$\overline{S} = \{x \in \mathbb{R}^n \,|\, x_n \to x, \text{ where } (x_n) \subseteq S \text{ is a convergent sequence}\}.$$

**Definition A.18.** Let $S \subseteq \mathbb{R}^n$. Then the interior $\text{int}(S)$ of $S$ is

$$\text{int}(S) = \mathbb{R}^n \setminus \overline{\mathbb{R}^n \setminus S}.$$

The boundary $\partial S$ is

$$\partial S = \overline{S} \cap \overline{\mathbb{R}^n \setminus S}.$$

**Definition A.19.** A function

$$f : S \to \mathbb{R}^n,$$

where $S \subseteq \mathbb{R}^m$ is called continuous if $f(x_n) \to f(x)$ for every convergent sequence $(x_n) \subseteq S$ with $x_n \to x \in S$.

**Corollary A.23.** *Let $f : K \to \mathbb{R}$ be a continuous function, where $K \subseteq \mathbb{R}^n$ is a compact set. Then $f(K)$ is bounded and there exists $x, y \in K$ with*

$$f(x) = \inf\{f(z) \mid z \in K\}$$
$$f(y) = \sup\{f(z) \mid z \in K\}.$$

## A.2   Exercises and solutions

**Exercise A.1.** Use induction to prove the formula in [U.C., (A.2)].

**Solution A.1.** The base case in [U.C., (A.1)] is verified by straightforward calculations. The induction hypothesis is that

$$(x_1^2 + \cdots + x_{n-1}^2)(y_1^2 + \cdots + y_{n-1}^2) - (x_1y_1 + \cdots + x_{n-1}y_{n-1})^2$$
$$= (x_1y_2 - x_2y_1)^2 + \cdots + (x_{n-2}y_{n-1} - x_{n-1}y_{n-2})^2 \quad \text{(A.1)}$$

where the last sum is over the $2 \times 2$ minors in the matrix

$$\begin{pmatrix} x_1 & x_2 & \cdots & x_{n-1} \\ y_1 & y_2 & \cdots & y_{n-1} \end{pmatrix}.$$

We obtain the additive decomposition

$$(x_1^2 + \cdots + x_n^2)(y_1^2 + \cdots + y_n^2) - (x_1y_1 + \cdots + x_ny_n)^2 =$$
$$\left[(x_1^2 + \cdots + x_{n-1}^2)(y_1^2 + \cdots + y_{n-1}^2) - (x_1y_1 + \cdots + x_{n-1}y_{n-1})^2\right] +$$
$$\left[(x_1^2 + \cdots + x_{n-1}^2)y_n^2 + (y_1^2 + \cdots + y_{n-1}^2)x_n^2 - 2(x_1y_1 + \cdots + x_{n-1}y_{n-1})x_ny_n\right].$$

We may rewrite the first of the above terms as stated in (A.1). The second term may be written as

$$(x_1^2 + \cdots + x_{n-1}^2)y_n^2 + (y_1^2 + \cdots + y_{n-1}^2)x_n^2 - 2(x_1y_1 + \cdots + x_{n-1}y_{n-1})x_ny_n$$
$$= (x_1y_n - x_ny_1)^2 + \cdots + (x_{n-1}y_n - x_ny_{n-1})^2$$

which is exactly the extra $n$ minors we get from adding the column $(x_n, y_n)$.

$$\star \quad \star \quad \star$$

**Exercise A.2.**

(i) Show that
$$2ab \le a^2 + b^2$$
for $a, b \in \mathbb{R}$.

(ii) Let $x, y \in \mathbb{R}^n \setminus \{0\}$, where $x = (x_1, \ldots, x_n)$ and $y = (y_1, \ldots, y_n)$. Prove that
$$2 \frac{x_i}{|x|} \frac{y_i}{|y|} \le \frac{x_i^2}{|x|^2} + \frac{y_i^2}{|y|^2}$$
for $i = 1, \ldots, n$.

(iii) Deduce the Cauchy-Schwarz inequality from (ii).

**Solution A.2.** (i) Notice that $a^2 + b^2 - 2ab = (a - b)^2 \ge 0$.

(ii) Using (i) with $a = x_i/|x|$ and $b = y_i/|y|$, we obtain
$$2 \frac{x_i}{|x|} \frac{y_i}{|y|} \le \frac{x_i^2}{|x|^2} + \frac{y_i^2}{|y|^2}.$$

(iii) If $x = 0$ or $y = 0$, the Cauchy-Schwarz inequality holds. Otherwise, applying the inequality from (ii) for $i = 1, \ldots, n$ we establish that
$$2 \frac{x^t y}{|x||y|} = 2 \frac{x_1 y_1 + \cdots + x_n y_n}{|x||y|} \le \frac{x_1^2 + \cdots + x_n^2}{|x|^2} + \frac{y_1^2 + \cdots + y_n^2}{|y|^2} = 2$$
or, equivalently, $x^t y \le |x||y|$. The inequality holds for $-x$ instead of $x$ as well and consequently, we get the result.

$$\star \quad \star \quad \star$$

**Exercise A.3.** Show formally that $1, 2, 3, \ldots$ does not have a convergent subsequence. Can you have a convergent subsequence of a non-convergent sequence?

**Solution A.3.** We consider the sequence $(n)_{n \in \mathbb{N}}$ and choose an arbitrary subsequence $(n_i)_{i \in \mathbb{N}}$, where $n_1 < n_2 < \cdots$. We will now show that this sequence do not converge to any point in $\mathbb{R}$ meaning that
$$\forall x \in \mathbb{R} \; \exists \epsilon > 0 \; \forall N \in \mathbb{N} \; \exists i \ge N : \; |n_i - x| > \epsilon.$$

Now let $x \in \mathbb{R}$, $\epsilon > 0$, and $N \in \mathbb{N}$. Since $x$ is fixed, we may choose $i \in \mathbb{N}$ such that $i > \max\{x + \epsilon, N\}$. Using this together with the fact that $n_i \ge i$ gives that
$$|n_i - x| \ge i - x > \epsilon.$$

The sequence $((-1)^n)_{n \in \mathbb{N}}$ does not converge but has the convergent subsequence $((-1)^{2n})_{n \in \mathbb{N}}$.

$$\star \quad \star \quad \star$$

**Exercise A.4.** Prove Proposition A.9.

**Solution A.4.**     (i) Let $\epsilon > 0$ and choose $N \in \mathbb{N}$ such that $|x_n - x| \leq \epsilon/2$
and $|x_n - x'| \leq \epsilon/2$ for every $n \geq N$. We may then use Theorem A.2
to write

$$|x - x'| = |x - x_n + x_n - x'| \leq |x - x_n| + |x' - x_n| \leq \epsilon$$

for every $n \geq N$. Since the left-hand side is independent of $n$, and $\epsilon$
was chosen arbitrarily, we conclude that $x = x'$.

(ii) For a given $\epsilon > 0$, let $N \in \mathbb{N}$ be such that $|x_n - x| \leq \epsilon/2$ and
$|y_n - y| \leq \epsilon/2$ for every $n \geq N$. Theorem A.2 gives that

$$|x_n + y_n - (x + y)| \leq |x_n - x| + |y_n - y| \leq \epsilon$$

for every $n \geq N$ which implies that $x_n + y_n \to x + y$.

To show that $x_n y_n \to xy$, we note the convenient identity

$$x_n y_n - xy = x(y_n - y) + y(x_n - x) + (x_n - x)(y_n - y).$$

This, together with Theorem A.2, implies

$$|x_n y_n - xy| \leq |x||y_n - y| + |y||x_n - x| + |x_n - x||y_n - y|$$

which becomes arbitrarily small for $n$ large enough.

$$\star \quad \star \quad \star$$

**Exercise A.5.** Let $S$ be a subset of the rational numbers $\mathbb{Q}$, which is
bounded from above. Of course this subset always has a supremum in $\mathbb{R}$.
Can you give an example of such an $S$, where $\sup(S) \notin \mathbb{Q}$?

**Solution A.5.** Consider

$$S = \left\{ \sum_{i=1}^{n} \frac{1}{i!} \,\middle|\, n \in \mathbb{N} \right\} \subseteq \mathbb{Q}.$$

It is well-known that $\sum_{i=1}^{n} \frac{1}{i!} \to e$ as $n \to \infty$ and, since $\left( \sum_{i=1}^{n} \frac{1}{i!} \right)_{n \in \mathbb{N}}$ is an
increasing sequence, it follows from Theorem A.10 that $\sum_{i=1}^{n} \frac{1}{i!} \to \sup(S)$
as $n \to \infty$. Now Proposition A.9(1) gives that $\sup(S) = e \notin \mathbb{Q}$.

$$\star \quad \star \quad \star$$

**Exercise A.6.** Let $S = \mathbb{R} \setminus \{0, 1\}$. Prove that $S$ is not closed. What is $\overline{S}$?

**Solution A.6.** Since $(1/n)_{n \geq 2}$ converges to $0$, $S$ is not closed. Furthermore, $(1 - 1/n)_{n \geq 2}$ converges to $1$ so by Definition A.15, we conclude that $\overline{S} = \mathbb{R}$.

$$\star \quad \star \quad \star$$

**Exercise A.7.** Let $S_1 = \{x \in \mathbb{R} \,|\, 0 \leq x \leq 1\}$. What is $\text{int}(S_1)$ and $\partial S_1$?
Let $S_2 = \{(x, y) \in \mathbb{R}^2 \,|\, 0 \leq x \leq 1, \ y = 0\}$. What is $\text{int}(S_2)$ and $\partial S_2$?

**Solution A.7.** From Definition A.18 we have that

$$\text{int}(S_1) = \mathbb{R} \setminus \{x \in \mathbb{R} \mid x \leq 0 \text{ or } x \geq 1\} = \{x \in \mathbb{R} \mid 0 < x < 1\}$$

and

$$\partial S_1 = \{x \in \mathbb{R} \mid 0 \leq x \leq 1\} \cap \{x \in \mathbb{R} \mid x \leq 0 \text{ or } x \geq 1\} = \{0, 1\}.$$

Furthermore, we find

$$\mathbb{R}^2 \setminus S_2 = \{(x, y) \in \mathbb{R}^2 \mid x < 0 \text{ or } x > 1\} \cup \{(x, y) \in \mathbb{R}^2 \mid y \neq 0\}$$

implying

$$\text{int}(S_2) = \mathbb{R}^2 \setminus \overline{\mathbb{R}^2 \setminus S_2} = \emptyset.$$

Finally, we conclude that

$$\partial S_2 = \{(x, y) \in \mathbb{R}^2 \mid 0 \leq x \leq 1, \ y = 0\}.$$

$$\star \quad \star \quad \star$$

**Exercise A.8.** Let $S \subseteq \mathbb{R}^n$. Show that $\text{int}(S) \subseteq S$ and $S \cup \partial S = \overline{S}$. Is $\partial S$ contained in $S$?
Let $U = \mathbb{R}^n \setminus F$, where $F \subseteq \mathbb{R}^n$ is a closed set. Show that $\text{int}(U) = U$ and $\partial U \cap U = \emptyset$.

**Solution A.8.** If we suppose that $x \in \text{int}(S)$, we have $x \notin \overline{\mathbb{R}^n \setminus S} \supseteq \mathbb{R}^n \setminus S$ which implies that $x \in S$.

Note that $S, \partial S \subseteq \overline{S}$ gives $S \cup \partial S \subseteq \overline{S}$. On the other hand we have $\overline{S} \cap \overline{\mathbb{R}^n \setminus S} = \partial S$ and $\overline{S} \cap (\mathbb{R}^n \setminus \overline{\mathbb{R}^n \setminus S}) \subseteq S$, and this implies $\overline{S} \subseteq S \cup \partial S$. It does not necessarily hold that $\partial S \subseteq S$ as this would imply that $S$ is closed by the relation $S \cup \partial S = \overline{S}$ and Proposition A.16.

For the open set $U$ we have that

$$\text{int}(U) = \mathbb{R}^n \setminus \overline{\mathbb{R}^n \setminus U} = \mathbb{R}^n \setminus \overline{F} = \mathbb{R}^n \setminus F = U$$

and

$$\partial U \cap U = \overline{U} \cap \overline{\mathbb{R}^n \setminus U} \cap U = \overline{F} \cap U = F \cap U = \emptyset.$$

⋆ ⋆ ⋆

**Exercise A.9.** Show that

$$\big||x| - |y|\big| \le |x - y|$$

for every $x, y \in \mathbb{R}^n$.

**Solution A.9.** It follows from Theorem A.2 that

$$|x| = |x - y + y| \le |x - y| + |y|$$

or equivalently,

$$|x| - |y| \le |x - y|.$$

The same inequality holds for $x$ and $y$ interchanged, and the result follows.

⋆ ⋆ ⋆

**Exercise A.10.** Give an example of a subset $S \subseteq \mathbb{R}$ and a continuous function $f : S \to \mathbb{R}$, such that $f(S)$ is not bounded.

**Solution A.10.** By choosing $S = \mathbb{R}$ and $f : S \to \mathbb{R}$ as $f(x) = x$ we have that $f(S) = \mathbb{R}$, which is indeed not bounded. If we insist that $S$ is bounded, choose $S = (0, 1)$ and $f : S \to \mathbb{R}$ as $f(x) = 1/x$. Then $f(S) = (1, \infty)$, which is unbounded. By Theorem A.22 we cannot find such an example if $S$ is both closed and bounded.

# Appendix B

# Linear (in)dependence and the rank of a matrix

## B.1 Introduction

The following exercises aim to acquaint the reader with linear dependence of vectors and the rank of a matrix; two fundamental concepts in the main material.

**Definition B.1.** A set of vectors $V = \{v_1, \ldots, v_r\} \subseteq \mathbb{R}^n$ is called linearly dependent if there exists $\lambda_1, \ldots, \lambda_r \in \mathbb{R}$ not all zero, such that

$$\lambda_1 v_1 + \cdots + \lambda_r v_r = 0.$$

Similarly $V$ is (or $v_1, \ldots, v_r$ are) called linearly independent if $V$ is not linearly dependent.

**Theorem B.2.** *The system*

$$a_{11} x_1 + \cdots + a_{1n} x_n = 0$$
$$a_{21} x_1 + \cdots + a_{2n} x_n = 0$$
$$\vdots$$
$$a_{m1} x_1 + \cdots + a_{mn} x_n = 0$$

(B.1)

*of linear equations always has a non-zero solution if $m < n$.*

**Theorem B.9.** *Let $A$ be an $m \times n$ matrix of rank $n$. Suppose that $r < n$ and that the first $r$ rows $A_1, \ldots, A_r \in \mathbb{R}^n$ of $A$ are linearly independent.*

(1) *There exists $r < j \leq n$, such that $A_1, \ldots, A_r, A_j$ are linearly independent.*

(2) *With the notation from (1) there exists $x \in \mathbb{R}^n$ with $A_1 x = \cdots = A_r x = 0$ and $A_j x \neq 0$.*

(3) *If $r = n - 1$, then any $x \in \mathbb{R}^n$ with $A_1 x = \cdots = A_r x = 0$ is uniquely determined up to multiplication by a constant.*

**Lemma B.6.** *Consider an $m \times n$ matrix $A$ with rows $A_1, \ldots, A_m \in \mathbb{R}^n$. If the first $s$ rows $A_1, \ldots, A_s$ are linearly independent and $A_1, \ldots, A_s, A_j$ are linearly dependent for every $j > s$, then for every row $A_i$ there exists $\lambda_1, \ldots, \lambda_s \in \mathbb{R}$ such that*

$$A_i = \lambda_1 A_1 + \cdots + \lambda_s A_s \qquad \text{(B.2)}$$

*and* $\operatorname{rank}(A) = s$.

## B.2   Exercises and solutions

**Exercise B.1.** Find $\lambda_1, \lambda_2, \lambda_3 \in \mathbb{R}$ not all $0$ with

$$\lambda_1 \binom{1}{2} + \lambda_2 \binom{3}{4} + \lambda_3 \binom{5}{6} = \binom{0}{0}.$$

**Solution B.1.** First, it should be noted that Theorem B.2 tells us that it is possible to find a non-zero solution as we have two equations in three variables. For instance, by imposing that $\lambda_1 = 1$ we know that

$$\begin{pmatrix} 3 & 5 \\ 4 & 6 \end{pmatrix} \begin{pmatrix} \lambda_2 \\ \lambda_3 \end{pmatrix} = \begin{pmatrix} -1 \\ -2 \end{pmatrix}$$

and this means $(\lambda_2, \lambda_3) = (-2, 1)$.

$$\star \quad \star \quad \star$$

**Exercise B.2.** Show that a non-zero solution $(x, y, z)$ to [U.C., (B.1)] must have $x \neq 0, y \neq 0$ and $z \neq 0$. Is it possible to find $\lambda_1, \lambda_2, \lambda_3$ in Exercise B.1, where one of $\lambda_1, \lambda_2$ or $\lambda_3$ is $0$?

**Solution B.2.** We are considering the system

$$\begin{pmatrix} 2 & 1 & -1 \\ 1 & 1 & 1 \end{pmatrix} \begin{pmatrix} x \\ y \\ z \end{pmatrix} = \begin{pmatrix} 0 \\ 0 \end{pmatrix}. \qquad \text{(B.3)}$$

Since the vectors $(2, 1)$, $(1, 1)$, and $(-1, 1)$ are pairwise linear independent, any solution to (B.3) with one zero entry must have every entry equal to zero. For instance, consider a solution $(x_0, y_0, 0)$. Then (B.3) shows that $(x_0, y_0)$ satisfies

$$\begin{pmatrix} 2 & 1 \\ 1 & 1 \end{pmatrix} \begin{pmatrix} x_0 \\ y_0 \end{pmatrix} = \begin{pmatrix} 0 \\ 0 \end{pmatrix},$$

and this implies $(x_0, y_0) = (0, 0)$ since the associated matrix is invertible.

The system from Exercise B.1 also consists of pairwise linear independent vectors so we have the same conclusion: if $\lambda_1$, $\lambda_2$, or $\lambda_3$ equal zero then all of them will be zero.

$$\star \quad \star \quad \star$$

**Exercise B.3.** Can you find a non-zero solution to

$$x + y + z = 0$$
$$x - y + z = 0,$$

where

   (i) $x = 0$?
   (ii) $y = 0$?
   (iii) $z = 0$?
   (iv) What can you say in general about a system

$$ax + by + cz = 0$$
$$a'x + b'y + c'z = 0$$

of linear equations in $x, y$ and $z$, where a non-zero solution always has $x \neq 0, y \neq 0$ and $z \neq 0$?

**Solution B.3.**    (i) When $x = 0$, the remaining system of equations only has the solution $y = z = 0$.

  (ii) When $y = 0$, the associated $2 \times 2$-matrix is singular (the reduced system has infinitely many solutions), and in particular we may find a non-zero solution. Specifically, any $(x, z) \in \mathbb{R}^2$ with $x = -z$ is a solution.

 (iii) This situation is similar to (i).

 (iv) The statement that every non-zero solution only has non-zero components is the same as saying that a solution with one component equal to zero implies that it is the trivial solution. Consequently, we know that if we set any of the components equal to zero, the associated matrix to the reduced system is invertible. This means that the three pairs $(a, b)$ and $(a', b')$, $(a, c)$ and $(a', c')$, and $(b, c)$ and $(b', c')$ are linearly independent.

$$\star \quad \star \quad \star$$

**Exercise B.4.** Check carefully that

$$\left( \tfrac{1}{a_{11}}(-a_{12}a_2 - \cdots - a_{1n}a_n), a_2, \ldots, a_n \right)$$

really is a non-zero solution to (B.2) in the proof of Theorem B.2.

**Solution B.4.** Using the induction hypothesis in [U.C., (B.3)], we find

$$\tfrac{1}{a_{11}}(-a_{12}a_2 - \cdots - a_{1n}a_n)a_{i1} + a_2 a_{i2} + \cdots + a_n a_{in}$$
$$= (a_{i2} - \tfrac{a_{i1}}{a_{11}} a_{12})a_2 + \cdots + (a_{in} - \tfrac{a_{i1}}{a_{11}} a_{1n})a_n$$
$$= 0$$

for $i = 2, \ldots, m$. Furthermore, we verify that

$$\tfrac{1}{a_{11}}(-a_{12}a_2 - \cdots - a_{1n}a_n)a_{11} + a_2 a_{12} + \cdots + a_n a_{1n} = 0.$$

Finally, we note that the solution is non-zero by the induction hypothesis.

$$\star \quad \star \quad \star$$

**Exercise B.5.** Compute the rank of the matrix

$$\begin{pmatrix} 1 & 2 & 3 \\ 4 & 5 & 6 \\ 14 & 19 & 24 \\ 6 & 9 & 12 \end{pmatrix}.$$

**Solution B.5.** Initially we note that $(1, 2, 3)$ and $(4, 5, 6)$ are linearly independent. Furthermore, $2(1, 2, 3) + (4, 5, 6) = (6, 9, 12)$ and $2(1, 2, 3) + 3(4, 5, 6) = (14, 19, 24)$, and Lemma B.6 implies that the matrix has rank 2.

Printed in the United States
By Bookmasters